UNLV Business and Technical Writing

Third Edition

JEFFREY JABLONSKI
University of Nevada—Las Vegas

KENDALL/HUNT PUBLISHING COMPANY
4050 Westmark Drive   Dubuque, Iowa 52002

First and second edition was a Kendall/Hunt website component.
Textbook added to this third edition.

Copyright © 2008 by Kendall/Hunt Publishing Company

ISBN 978-0-7575-4924-3

All rights reserved. No part of this publication may be reproduced,
stored in a retrieval system, or transmitted, in any form or by any means,
electronic, mechanical, photocopying, recording, or otherwise,
without the prior written permission of the copyright owner.

Printed in the United States of America
10 9 8 7 6 5 4 3 2 1

Table of Contents

Preface — ix
Acknowledgments — xi

I. Principles

CHAPTER 1—*Business Writing* — 3
    Good Communication Skills Needed to Get Ahead in Business — 3
    No Microwave Food—Ever! — 4
    The Fundamental Principles for Effective Business Writing — 6
    Conclusion: "Good Writing" Is More Than "Good Grammar" — 11
    Exercises — 13

CHAPTER 2—*Technical Writing* — 15
    Why Study Technical Writing? — 16
    Effective Technical Writing Is Rhetorical — 17
    Conclusion — 21
    Exercises — 22

CHAPTER 3—*Professional Writing Style* — 23
    Revision versus Editing — 23
    Principles of Effective Professional Writing — 24
    Draft, Revise, Take a Break…Then Revise Again — 30
    Exercises — 32

CHAPTER 4—*Business Correspondence* — 35
    Memos — 35
    Letters — 37
    E-mail — 38
    Generic Structure of Memos, Letters, and E-Mail — 39
    Correspondence Style — 42
    Common Genres of Correspondence — 43
    Writing Electronic Correspondence — 46
    Exercises — 48

CHAPTER 5—*Reports* — 51
    Reports as a Type of Professional Writing — 51
    Report Format — 53
    Report Design — 56

Graphics	57
Adopt a Style Guide	64

CHAPTER 6—*Resumes* — 67

Purposes	67
Format	68
Organization	69
Content	75
Style	77
Resume Visual Design	78
Scannable Resumes	80
Conclusion: Do it Your Way!	82

CHAPTER 7—*Cover Letters* — 83

Aims of the Cover Letter	83
Cover Letter Format	85
Should You Submit an E-mail Cover Letter?	88

CHAPTER 8—*Presentations* — 91

Purpose	91
Audience	92
Organization	92
Visual Aids	93
Delivery	97
The Big Day	99

CHAPTER 9—*Collaboration* — 101

Developing Team Skills	101
Team Leadership and Roles	103
Organizing and Managing the Project	104
Document Production	106
Revision and Editing Stages	109

CHAPTER 10—*Definitions* — 113

Types of Definitions	113
Writing Effective Technical Definitions	115
Potential Problem Areas	115

CHAPTER 11—*Usability Testing* — 119

How are Usability Tests Conducted?	119

II. Projects

CHAPTER 12—*Introductory Memo Project* … 127
- Steps for Completing this Assignment … 127
- Follow the Required Memo Format … 128
- Write Using a Business Style … 128
- Submission Format … 129

CHAPTER 13—*Situation Analysis* … 131
- Preliminary Considerations … 132
- Audience … 132
- Purpose and Intended Use(s) … 133
- The Writer … 133
- Document Design … 134
- Notes … 134
- Optional: Situation Analysis Memo … 135

CHAPTER 14—*Project Assessment Memo* … 137
- Overview … 137
- Context … 137
- Documents … 138
- Production … 138
- Summary … 138

CHAPTER 15—*Definitions Project* … 139
- Background: Bad Culture – New Management – New Policies … 139
- Exercises … 141

CHAPTER 16—*Cases: Overview of Goals and Strategies* … 143
- Questions for Analyzing Cases … 144
- Focus on Managing Ethical Dilemmas … 145

CHAPTER 17—*Big-1 Rental Agency Case* … 147
- Exercises … 150

CHAPTER 18—*A Business Faux Pas Case* … 153
- Jeaneaux Letter … 156
- Nester Letter … 158
- Exercises … 159

CHAPTER 19—*The Scanner Slip-Up Case* … 161
- Official Rules: Thomson's Sales Training Program … 164

E-Mail Winning Notification	165
Exercises	166

CHAPTER 20—*Insurance Fraud at MedTech Case* — 167

Letter from Susan Seer	169
Susan Seer's physician's report/certification of disability	170
Online Resources	171
Exercises	172

CHAPTER 21—*Foodborne Illness on Festival Case* — 173

Exercises	175

CHAPTER 22—*Job Search Project* — 177

Find an Actual Job Advertisement	179
Research the Company	182
Job Analysis Memo	184
Resume and Cover Letter	186
Project Assessment Memo	188

CHAPTER 23—*Staff Development* — 191

Deliverables	192
Planning	194
Proposal	195

CHAPTER 24—*Procedures Project* — 203

Deliverables	204
Steps for Completing the Project	205

CHAPTER 25—*Client Project* — 209

Deliverables	210
Planning	211
Project Plan	220
Client Letter	222
E-Mail Progress Report	223
Final Report	224
Project Assessment Memo	231

CHAPTER 26—*Usability Project* — 233

What Does Usability Have to do with Learning How to Write?	234
Deliverables	235
Choosing a Client Web Site	235
Design Plan Report	236

Progress Report	240
Usability Report	241
Project Assessment Memo	244

CHAPTER 27—*International Project* — 245

Deliverables	246
Planning	247
General Evaluation Criteria for the International Project	249

Preface

The staff who teach business and technical writing at UNLV want to thank you for purchasing *Business and Technical Writing*. We hope you find this text informative, accessible, and useful. The text you will be using in this course is written by UNLV faculty for UNLV students. It is based on research-tested best practices nationwide. However, this text has in mind the needs and abilities of UNLV students. By writing a text just for you and your fellow UNLV students we didn't have to worry about how a teacher on the east coast, who is at some very different academic institution with very different kinds of students, believes business writing should be taught. So, basically, we could focus on your needs, making the text more relevant and, importantly, keeping the cost more affordable.

This text and the accompanying projects aim to develop multiple business literacies: rhetorical, visual, information, computer, and ethical. We believe multiple literacies are necessary for interacting with the more sophisticated information mediums used in business and technical settings today. It is in the development of these multiple literacies that you will grow in sophistication as a critical thinker and as a business or technical writer. Foremost, this text will develop your rhetorical skills: your ability to analyze the demands of a given workplace writing situation and adapt your writing given your understanding of purpose and audience. This text also emphasizes the role of visual design in effective workplace writing. You will learn how page layout, graphics, text (and hypertext) operate rhetorically and contribute to clear and effective communication. Furthermore, this text aims to develop your information literacy, or your ability to find the necessary information and integrate effectively into your own writing without plagiarizing. The text also will develop your ability to use the computer as a tool for communication, including how to compose documents electronically and how to use the World Wide Web for finding and retrieving information. This text also emphasizes the ethics of workplace communication, that good writing is not only mechanically sound but also maintains positive relations and avoids harming others.

This textbook also aims to support the UNLV Business and Technical Writing Program's project-based approach. All of our UNLV business and technical writing courses integrate project-based learning. Traditionally, learning has all too often been passive; the teacher lectures about some subject matter, and the students passively take it all in. In traditional courses you're expected to memorize the material and are tested mostly on your ability to recall information rather than your ability to apply abstract concepts to concrete problems. Problem-based learning is much more active. Problem-based learning gives you a problem that requires knowledge and skill in certain principles to solve effectively. UNLV's business and technical writing courses are structured around a series of projects that require you to apply related principles of effective written communication. To solve the problems inherent in each project you will need to do more than just memorize principles of business and technical writing; you will need to apply them. In this way, information—the subject matter of the course—becomes more functional and learner-controlled. Research shows this deepens your learning and increases the chance you'll internalize the strategies you used, so you will better prepared to apply principles learned to similar situations in the future as you write in the workplace.

Your business and technical writing instructors hope that this textbook helps you develop more fully your business or technical writing skills and that your semester is a successful one.

Jeffrey Jablonski

Acknowledgments

Many current and former UNLV business and technical writing faculty and staff have contributed to this textbook in various ways, including writing content, designing and testing assignments, and researching background information and links.

Elyse Arring

Jenny Bania

Suszanne Berfalk

Jeff Burbank

Joseph Cameron

Don Capp

Rebecca Colbert

Jason Coley

Anish Dave

Paris Drebes

Jaq Greenspun

Vanessa Huff

Chad Leitz

Karen Lovat

Heather Lusty

Ed Nagelhout

Michael McCombs

Cory Ness

Jonathan Peters

Constance Pruss

Lara Ramsey

Kelle Schillaci

Homer Simms

Julie Staggers

Paul Straus

Denise Tillery

Jenny Toups

PART I

Principles

Chapter 1
Business Writing

It is no secret that strong written communication skills are essential for business success. That's why a course in business writing is required for most of you reading this: your advisors do not want to leave it up to chance that you would *elect* to take a course that's focused on writing. To help you see just how important and valuable strong business writing skills are, this chapter discusses the following:

- Facts that demonstrate the importance of strong business writing skills
- An example of bad business writing
- The fundamental principles of business writing stressed in this course

Good Communication Skills Needed to Get Ahead in Business

Most employers complain that their employees can't write. Thirty percent of all business memos and e-mails are written solely to clarify earlier written communication that didn't make sense. This translates into an estimated $225 billion a year in lost productivity.[1] The rise of the Internet and the use of e-mail has increased the need for strong writing skills, as many companies are switching to e-mail as the main means of communication within the company and to outside customers.

Survey after survey indicates that managers believe their employees lack the basic writing skills needed in today's information economy.[2,3,4,5] A widely publicized 2004 report titled *Writing: A Ticket to Work...Or a Ticket Out* found that workplace writing is viewed by managers as a threshold skill used to make hiring and promotion decisions.[6] This survey of 120 major American companies concluded the following:

- Employees "frequently" to "almost always" produce memos, letters, and reports, and communication through e-mail and PowerPoint presentations is "almost universal"[6]
- People with weak communication and writing skills will not be hired and are more likely to be fired than promoted[6]
- The large majority of companies (80%) assess writing ability during the hiring process
- Many companies (50%) take writing into account when making promotion decisions

- American companies spend over 3 billion annually training employees with weak writing skills

The above report was produced by the National Commission on Writing, an organization formed in 2002 to address the "growing concern within the education, business, and policy-making communities that the level of writing in the United States is not what it should be." In 2005 the Commission issued a follow-up report based on a survey of government sector managers titled *Writing: A Powerful Message from State Government*. The report amplified the findings of the previous survey, claiming that writing was an even more important consideration in government work, given the need for state employees to clearly communicate information about laws, regulations, and policies to citizens.[7]

Not convinced yet that it's important to have strong business writing skills? Consider these statistics:

- Almost 90% of business school graduates credited their writing skills as helping to accelerate their advancement[2]
- Almost 80% of upper level managers identified business writing as the most useful college course they had taken[2]

To prepare you for writing in the workplace, this textbook emphasizes the ways that business writing differs from the way you've been taught to write for school. For instance, you'll learn that, unlike college essays, it is preferable to use first person "I" and second person "you" voice in most business writing. Unlike college essays, you should write memos and letters using single-spacing and shorter paragraphs. You should use headings and other visual elements like bulleted lists and graphic aids. To suit busy readers such as executives and customers, your business writing should generally be direct and to the point, unlike a lot of college writing. And, you may or may not be heartened to learn this, you will not be rewarded in business writing for using big words and fluff filler—the type of writing some professors may have encouraged in your past college experience. This habit of "academic-ese" that college graduates bring to their workplace writing is a big reason employers complain that employees can't write.

This textbook asks you do a lot of writing. Lack of practice is another reason people complain that no one can write anymore. Like any skill, writing is an ability that atrophies without use, and students are rarely asked to write in school. The more you write, the better your writing becomes.

The next section discusses an example of bad business writing to make the point that good business writing involves more than attention to surface features of grammar and mechanics.

No Microwave Food—Ever!

What do you think of the following memorandum (based on an actual memo, written at a real company)

> From: Standish, Royce (Operations, Acme Widgets Inc.)
> Sent: Wednesday, February 18, 2004
> **Subject: Microwave policy**
>
> Attention everyone: I have nothing against warm food...However, from here on out if you must reheat food you must do so in a microwave safe container. If you are found reheating food in the microwave using foil or other combustible materials you will be terminated, period! The reason for this is that most of all fire alarms in office buildings are because some moron set the microwave on fire with aluminum foil. There was another incident last night... There have been 4 alarms in the last 2 months. These microwave fires cost us between $5,000 and $8,000 each alarm. If you must have warm food; order out or use plastic, or you're fired. Thank you.
>
> Royce Standish
> Vice President,
> Operations
> 1-800-xxx-xxxx Ext xxxxx

Laughable, right? But would you want to work at this company? What words would you use to describe Mr. Standish, the author of the memo? Would you want to work for Mr. Standish?

As with all business writing, the Fire Policy example serves many purposes, some explicit, others less obvious. The memo's main purpose is obviously to *inform* all employees of the ban on combustible materials in the microwave and the consequences for violating the new policy. Mr. Standish includes details about the costs associated with microwave fires to *persuade* employees to adhere to the policy. As another persuasive tactic Standish *invokes his authority* to terminate employees. This memo also has other, less explicit purposes. It *conveys an image* of the writer and *influences the culture* of the company. How did you describe Standish? Sarcastic, condescending, heavy-handed? Is that the kind of image a business executive should project? Do you want to work in a business environment where a vice president casually refers to his employees as idiots in writing? It makes one wonder what else he thinks of his employees.

The Fire Policy memo also functions as a *record* of Mr. Standish's new policy—and his communication style—within his company. Perhaps Mr. Standish regrets writing this memo. But it's too late. The memo was e-mailed to hundreds of employees.

One thing is likely: Mr. Standish never took a business writing class. While he may have risen to the rank of vice president, he certainly hasn't done it in a way that would endear himself to his employees. Maybe he's not even the best person for the job. Is a ban on "combustible materials" the best solution to the problem at this company? There's more wrong with the Fire Policy memo than just formalistic concerns of tone and punctuation.

Good business writing maintains positive relations and demonstrates careful, critical thinking. The next section will outline key principles that should help you avoid mistakes like those in the Fire Policy memo.

The Fundamental Principles for Effective Business Writing

Rhetoric is the ancient art of effective communication. Today, rhetoric all too often has negative connotations, as the word is sometimes used to describe deceptive or evasive language. But there is a more positive side to rhetoric. The ancients recognized that speaking and writing abilities could be improved through the study of effective communication and deliberate practice, which came to be known as the discipline of rhetoric.

Paying attention to rhetoric as you plan and compose your documents raises the following key questions:

- What is the *purpose* of my writing, or what am I trying to accomplish?
- Who is being addressed, or who is my *audience*?
- How does the *context*, or circumstances within which the communication occurs, affect what I write?
- What *image*, or impression of myself, do I want to create with my reader, what should the reader think of me?

Purpose, audience, context, image. As you prepare to write, the more you are aware of and account for what your purposes are, who your audience is, what context or situation you're writing in, and what image of yourself you want to convey, the more effective your written communication will be.

Know Your Purposes

The main goal of all business communication is to *develop and maintain positive relations between people*. Even if a sales letter doesn't make an immediate sale, that doesn't mean it failed—the letter helps develop a relationship with the customer. Clichés like "the customer is always right" are designed to help employees maintain positive business relations. Far too often, people forget this and communication dissolves into hostile, threatening, or condescending exchanges. There's a negative stereotype often associated with business. Portrayals of businessmen from Charles Dickens' Scrooge to Gordon Gekko, the greedy broker played by Michael Douglas in the movie *Wall Street*, reflect the belief that to be successful in business one must be ruthless and cutthroat. While some may have enough power or influence to act this way, few everyday business writers can afford to risk damaging relations with their readers, be they customers, clients, supervisors, or co-workers. And even if you could get away with heartlessness, is that really the way business *should* be done?

Other purposes important to business writing include:

- **To inform:** Most business writing communicates information to readers that will allow them to do something better, whether it be to act or think differently. The challenge is to understand what information is essential to helping readers act. It's also essential to include information about harms or risks that readers need to make informed decisions.[8]

- **To persuade:** Information alone rarely compels readers to act. They must be persuaded about what to do with the information. After asking "*what* is this information?" readers want to know "*why* should I act on this information?" Information frequently must include arguments, assertions, and opinions about its usefulness. Effective automobile advertisements make the connection between why good handling is desirable (to swerve out of the way of earthquake faults and falling boulders). Sales letters aren't the only occasion for persuasion. A worker's performance evaluation or progress report attempts to persuade a supervisor that she's being productive. When attempting to persuade, writers must be careful not to manipulate information for their own ends. Writers must be fair and honest and make sure communication is not limited to their own interests, but also considers the best interests of all parties involved.[8]

- **To establish a legal record:** Most writing in business *documents*, or creates a record, of the communication between two or more parties.[9] Writing in business is often legally binding and always has legal implications. Nearly every year there is a big news story that involves some "smoking memo" implicating a major corporation in wrongdoing of some sort. In 2000, Firestone Tires recalled 6.5 million tires because they were associated with numerous fatalities caused by Ford Explorer rollover accidents. Subsequent investigations revealed early memo reports written by engineers within Ford Motor Co. alerting management to design flaws contributing to the rollover problems. These memos suggested some negligence on the part of Ford. Does this mean that the engineers should not have written about the problems? Does it mean that Ford management should have destroyed the memos? No, the information is part of the ongoing work of the organization and serves as the organization's memory. Ford engineers and managers can retrieve memos like this and familiarize themselves with the history of the project. But so can lawyers and the public.

Pay attention to the record-keeping function of business writing and the reality that future, unforeseen readers have access to an organization's documents. Frequently, documents need additional background information to accommodate future readers. This is why you should pay attention to details such as dates and facts when you write. Always ask yourself if what you've written is something you want to put "in print." E-mail is frequently treated as an informal communication medium like speech, but high-profile lawsuits have demonstrated that e-mail is not a secure, private way of communicating. It is typically monitored by an organization, and even if deleted, it is typically backed up and recoverable.

Analyzing your purposes for writing also emphasizes that business writing is an *action* done to cause further action. Most students associate (school) writing with meaningless busywork. But writing is central to the activity and goals of business organizations. People write in business for many reasons:

- Announce a new or updated product or service
- Establish a new policy
- Return credit to clients
- Request information or meetings
- Complain about services or products
- Inform clients of rate increases
- Praise or compliment services or products
- Express thanks
- Respond to a complaint (called an "adjustment letter")
- Refuse requests[8]

Can you think of other purposes not listed here? Notice how each of the reasons listed above begins with an *action verb*: to *sell*, to *thank*, to *request*, etc. Before you write, brainstorm a list of purposes and think about which will be most important. The more you can articulate the aims of your writing, based on your analysis of the situation, the more your writing will improve because it will give you goals and a means to measure how well your draft accomplishes your goals. Keep in mind that in any typical business situation **there are multiple purposes** and some may even conflict. And always check that your writing doesn't risk damaging relations, creates a suitable written record, and doesn't misrepresent or mislead.

Understand Your Audience

Audience is the key factor that shapes texts. Who your audience is will determine the level of formality of the writing, what terms should be defined, what information should be included, and how that information should be arranged. You should also consider the social and organizational status of your audience. Is your audience superior to you, such as your boss, or are they inferior to you, such as employees you manage? Are they insiders or outsiders? Insiders might be company or professional peers who are more familiar with your subject and the language you want to use. Outsiders might be the public or clients with less knowledge about your topic who may require simpler language or need technical terms defined so that they can understand. Writers always have to make a choice about how much they can assume an audience knows about their topic. Other factors such as age, gender, class, race, nationality, occupation, and political affiliation influence the receptivity of an audience to a text. As you plan your writing, you should always try to shift your perspective from "what do I want to say?" to "what information does my audience need to understand my message?"[10]

As the discussion of business writing as a legal record suggests, when you write in business you will never have only one reader, even if you are addressing only one reader.[8] You must always consider how your writing accommodates important secondary readers, anyone who is not the primary or immediate addressee of the document. A new policy written by an HR manager may be addressed to an immediate supervisor for approval but have HR office staff in mind as its primary audience. The policy, because it affects employees, may need approval by lawyers. The HR manager may have written the policy with her professional peers in mind because she wants to ensure that her organization manages human resources in line with professional standards, but the policy must also be written for superiors, lawyers, and employees and managers who aren't even in the company yet. It must also be intelligible to fellow HR staff in the event that the HR manager is promoted or moves to another company.

Consider the Context in Which You Write In

Context as a rhetorical principle is a complex notion that includes an awareness of the immediate physical situation in which a person is writing. It also includes a more abstract situational awareness of other factors that influence the production and reception of text, including the organizational, cultural, historical, ethical, and legal contexts or perspectives that shape documents. Successful managers are often considered to possess a certain business savvy about what's feasible within an organization. This savvy reflects a high degree of contextual awareness and an ability to anticipate appropriate actions given the constraints of a particular situation. When writing, you should always try to view the outcomes of what you want to accomplish and how you want to accomplish it from a variety of perspectives.

In business writing, the organization's *corporate culture*, its shared values, norms, roles, rituals, and beliefs, shapes permissible action. Many writers get into trouble because they don't think through the implications of what they write. Corporate culture influences writers' choices of content, persuasive approach, and word choice. For instance, one writer was told to delete the word *hope* from a draft. "We don't *hope* for anything around here," she was advised, "we *decide* what we want and then we make it happen."[11] Many organizations have clear chains of command and lines of communication. Ignoring either of these may create serious problems for someone wanting to accomplish something in writing, be it to complain about something or to change a way of doing something. The corporate culture determines when and why who speaks to whom, and who listens to whom.

Individuals within companies also affect what's written. Powerful executives' preferences evolve into maxims for behavior within the company. External factors also affect business communication, including governments, media, markets, unions, regulatory bodies, suppliers, trade organizations, competitors, and the public. Often, what may seem permissible within the bounded context of the organization is unworkable in light of government standards, acceptable market practice, union regulations, or public opinion.[11]

More abstract factors such as the historical, cultural, and ethical contexts of communication are also important contextual considerations. Some action or argument may not be feasible because of recent events within an organization or larger society. After the dotcom bubble burst in the late 1990s, entrepreneurs had to change their tactics for raising money for business ventures. The Internet gold rush had ended and venture capitalists became much more skeptical about e-commerce business plans. Cultural considerations also affect communication. Since the civil rights movement of the 1960s more women and minorities have entered into the workplace, influencing all sorts of practices including acceptable communication. Now communication must be nonsexist and recognize diversity.

All communication must consider moral and ethical standards of conduct. What would you do if you were asked to contribute to a document that you knew could lead to harm of some sort for others? Would you falsify information if you knew you stood to gain a great deal of money? The impersonal bureaucracies of modern organizations give rise to ethical challenges that writers must confront. Everything you write is subject to the legal and ethical constraints of society. As James Porter notes, "Whenever you write, you have an obligation as a citizen and as a member of a community to write with the good of that community in mind. You are not just a worker in a business, you are also always a member of larger civic and social communities, whose interests you must keep in mind. 'I just did it because the boss told me' is not a valid defense if what you do is illegal or unethical."[9]

Establish an Appropriate Image

Everything you write within an organization contributes to readers' opinions about who you are—what kind of writer you are, what kind of thinker you are, and what kind of person you are. Frequently, people will *only* use what you write to form an opinion of you and your capabilities. Think about the job application process. Employers screen dozens to hundreds of resumes and cover letters submitted by hopeful job seekers, narrowing the pool down to a handful of applicants to interview in person. These employers make decisions about whether applicants can do the job based on what's written in the job application documents.

Ancient rhetoricians were conscious of how a person who registered a strong moral character and a sense of goodwill established strong *credibility*, or trustworthiness and believability, which could have a powerful persuasive influence on an audience. The Greeks termed the character or moral nature of a person *ethos*. Ethos is related to the image that writers project in their writing. The following elements of writing affect a writer's ethos, either negatively or positively:

- **Errors in a text:** Have you ever sent an e-mail to an employer or professor that your eighth grade English teacher would have marked up with a red pen? Readers tend to attach negative traits to writers of sloppy compositions, especially in business. Human resource managers will often stop reading resumes with just a few obvious errors.

- **Tone:** Customer service managers will respond more positively to an even-tempered complaint than a hostile one. In business you should always aspire to create a *professional ethos*. Do you want readers to see you as hostile, uncaring, and inattentive, or do you want readers to see you as even-tempered, considerate, and thorough?

- **Evidence used to support arguments:** A request for a raise would be more credible if the writer included specific details about the ways the writer improved productivity and sales within the past year. In academic writing, writers cite published sources to support their arguments and demonstrate to readers that they are familiar with what's already been written on a topic.

- **Status and reputation:** Right or wrong, a senior worker is more likely to be taken seriously than a new worker.

Another ethos-related factor to consider is that your business writing often represents not just your own views but those of the organization for which you write. The writing you do in business is often "ghostwritten," or not associated with your name. You may be asked to draft a letter for a supervisor's signature. You may write company promotional materials. You may be asked to write parts of documents that are primarily written by someone else or that represent the whole company. When you write as a representative of your workplace your identity is complicated: it overlaps with the identities of your organization and your profession. You can't just write what you want without regard to how it will reflect on the company to which you belong.

Your business writing simultaneously represents the viewpoints, values, concerns, and ethics of yourself and your organization and your profession. In this regard, your writing has to mediate those interests, which can conflict. Suppose you have to write a letter of apology to a client for the behavior of one of your staff members. You have obligations to yourself, your organization, and your client. You will want to apologize and restore the damaged relationship with your client without making your client feel wrong for complaining, without making yourself look overly apologetic, without making your employee look incompetent (for you hired him or her), and without making your company look desperate.[8]

Conclusion: "Good Writing" Is More Than "Good Grammar"

This chapter began by demonstrating the importance of written communication skills, particularly the value placed on such skills by employers. The second part presented a typical example of bad business writing to show that content is just as important as form in evaluating whether a document is well written. The last part outlined basic principles of rhetoric that allow writers to more carefully plan the content of their writing based on the rhetorical situation. This textbook aims to expand your notion of good writing beyond simply a command of the mechanics of writing. A facility with grammar, spelling, and punctuation is important, but it alone does not guarantee that a document is well written. In the complicated contexts of business writing, where the practical (read as "money"), legal, and ethical stakes are high, writers must be able to analyze and actively adjust their writing according to the purposes, audiences, contexts, and identities called for in each new writing situation.

There are other elements of good business writing not discussed in this introduction that this textbook helps develop. Effective writers also need *visual literacy*, the ability to recognize visual design as an element of written communication. Given the multimedia nature of the Internet and the advanced desktop publishing capabilities of word processing software, writers must be able to use page layout, typefaces, and graphics to make their documents more appealing and persuasive.

Writers also require *information literacy*, the understanding of how knowledge is constructed, organized, disseminated, and stored. Early visions of an Internet came about as theorists and scientists grappled with the idea of information overload, the realization that the rate of knowledge production was eclipsing our ability to process it. Effective business writers can find, analyze, and integrate information from a variety of sources (Internet, libraries, experts, surveys, users) into written products. Especially given the democratizing force of the Internet, another aspect of information literacy is the ability to evaluate the credibility, validity, and reliability of information. The temptation to copy and paste information from electronic sources is high; good business writers must know how to appropriately incorporate ideas, words, and images from a variety of sources into their writing.

Finally, writers must possess *ethical literacy*, the ability to understand and apply social and professional standards of appropriate action in everyday business situations. Steven Katz, a contemporary rhetorician, analyzed a memo written by a Nazi engineer during World War II according to traditional criteria for effective business writing. In terms of its *form* (its organization, style, tone, and argumentation) this memo is nearly perfect. However, the *content* of the memo is quite another story. It attempts to persuade a superior that technical changes need to be made to improve the efficiency of large trucks being used to kill prison camp detainees during the Final Solution, the Nazi's plan to exterminate European Jews and other so-called undesirables. The horrific subject matter of the memo was obfuscated by references to the "merchandise" and the "load."[12] This example underscores the importance of not reducing definitions of good writing to formalistic concerns of mechanics and style.

Whether they can put it into words or not, when employers complain that their workers can't write they are generally pointing to bigger problems than just spelling and punctuation; they are referring to writers' abilities to think critically about complex problems and respond appropriately in writing. These are the elements of writing constituted in the multiple literacies of business writing and emphasized in this course.

Exercises

1. Revise the Fire Policy Memo. Consider not just the language and tone but the appropriateness of the solution to the Fire Policy problem as well.

2. Individually or in a group, find a current, recent, or past event in the news that relates to business writing/communication, similar to the Ford/Firestone case discussed above. Use the Internet and your library's periodical indexes. Reflect on how the news story relates to issues discussed in this introduction.

3. Interview someone you know who writes in the workplace. Ideally, interview someone in the field you plan to enter. Brainstorm a list of interview questions in class, such as what types of documents are written, how much time is spent writing (including planning, drafting, and revising), and the type of training the writer received (in school and at work). Write up a profile of a business writer based on your interview and share it with the class.

4. Read the following online essay titled "The Death of Strunk and White" by Gary Henrickson, an English professor at North Dakota State College of Science. The essay can be found at: http://66.102.7.104/search?q=cache:mw4HB8sU9tEJ:mtprof.msun.edu/Fall1998/Henricks.html&hl=en&lr=&strip=1.

5. Write a response to Henrickson's essay. Your response might consider the following questions: Do you agree or disagree with Henrickson? Do you think Henrickson is being serious or satirical? What do you think of Henrickson's eight rules for real-life writing? Do any past experiences of your own correspond with or contradict Henrickson's points? If real-life business writing is so bad, what's the individual business writer's obligation to maintain standards of good writing as defined by this course.

End Notes

[1] "E-mail Exposes the Literacy Gap." *Workforce*. November 2002. p. 15.

[2] "Writing and Oral Communication Skills: Career Building Assets." Plain Language Network. http://www.plainlanguage.gov/Summit/writing.htm. Retrieved: 1 January 2003.

[3] "More than a Third of Applicants Lack Basic Skills." *HRFocus*. July 2000. p. 9.

[4] "Wanted: Leaders Who Can Lead and Write." *Workforce*. December 1997. p. 21.

[5] Fisher, Ann. "The High Cost of Living and Not Writing Well." *Fortune*. December 7, 1998. p. 244.

[6] National Commission on Writing. *Writing: A Ticket to Work…Or a Ticket Out, A Survey of Business Leaders.* September 2004. http://www.writingcommission.org/prod_downloads/writingcom/writing-ticket-to-work.pdf. Retrieved: 1 June 2006.

[7] National Commission on Writing. *Writing: A Powerful Message from State Government.* July 2005. http://www.writingcommission.org/prod_downloads/writingcom/powerful-message-from-state.pdf/. Retrieved: 1 June 2006.

[8] Toner, Lisa. "Writing Business Letters." Purdue Business Writing Coursepack. 1995. Lafayette, IN.

[9] Porter, James E. "What is Business Writing? Why Do It?" Purdue Business Writing Coursepack. 1994. Lafayette, IN.

[10] "Revision in Business Writing." Purdue Online Writing Lab handout. http://owl.english.purdue.edu/handouts/print/pw/p_revisebus.html. Retrieved 1 Jan 2003.

[11] Driskil, Linda. "Understanding the Writing Context in Organizations." In Kogen, Myra (Ed.), *Writing in the Business Professions.* Urbana, IL: National Council of Teachers of English. 1989. pp. 125–145.

[12] Katz, Steven. "The Ethic of Expediency: Classical Rhetoric, Technology, and the Holocaust." *College English.* Mar 92, Vol. 54, Issue 3, p. 255.

Chapter 2
Technical Writing

Technical writing is anything written that conveys highly specialized information. Specialized information is that which is related to a profession or academic discipline that has its own language for communicating the ideas and practices that are important to that field. To insiders—the engineers and architects, computer technicians, scientists, medical professionals, and lawyers that make up a field—this language is necessary for dealing with highly complex and abstract problems. To outsiders, this language, called *jargon*, is often difficult to understand.

Frequently, the cause of bad technical writing is that technical insiders fail to translate their specialized language for the benefit of outsiders with whom they are communicating. The most notorious examples of this are electronics manuals and software instructions, which are often written with such insider jargon as, "To initiate application initialization, populate the user ID field with the designated user identification authorization numeral." "What does that mean?" asks the befuddled couple trying to install the new accounting software they just purchased. What the technical writer should have written was, "To install the program, type the 15-digit number printed on page 2 of your manual into the box labeled 'Enter user ID here.'"

There are other reasons for bad technical writing, including lack of time, sloppiness, and deliberate obfuscation. Perhaps the writer of the example above didn't have time to check to see if that instruction made sense to an actual reader because it was part of a software upgrade that needed to be released in time for tax season. Or maybe the writer didn't check to see if the instruction made sense because he didn't care if anyone could actually use the product his or her company spent thousands of dollars developing. Or perhaps the writer deliberately wanted the statement to be so overly technical that users would have to hire a trained technician, likely from the company that produced the software, to install it for them.

Good technical writing effectively conveys information from a specialized field to specific readers, in specific situations, in such a way that the readers find the information useful. In this sense, *user friendliness* applies not only to computer software but to any document written for readers' instrumental use. Furthermore, good technical writing is that which is carefully produced, takes into consideration the needs and knowledge level of the reader, and doesn't deliberately misrepresent ideas for the writer's own benefit.

A common misconception is that science, engineering, and technology fields are neutral and that people who communicate in technical professions must be as neutral and objective as possible in their communications. Far from being neutral, technical professional activity occurs within social, cultural, historical, political, ethical, and other contexts. Communication never occurs in a vacuum. Whenever writers consciously (or unconsciously) adjust their message to the constraints of a given situation—and all writers must, to be effective—they are using rhetoric, the art of speaking or writing well. Techniques for making written information useful to actual readers include global, big picture issues such as understanding what readers will need to do with the information and adjusting the message accordingly, as well as micro, surface-level issues such as formatting the document with lots of spacing and easy-to-see section headings. This course will focus on these and many more techniques for effectively communicating technical information.

Why Study Technical Writing?

You may have heard we are in the information age. This means that ideas are considered tools—things that make our lives easier or help us get things done—that are just as important as the machines that replaced human labor in the industrial age. In other words, creating a database that can retrieve specific digital images captured during a continuous remote sensing experiment (e.g., a satellite that continuously takes pictures of a semi-dormant volcano's thermal emissions) is just as important as creating the digital equipment that generates the thousands of volcano images each day. While production jobs, like factory worker, will always be necessary, it is the symbolic-analytic jobs where people take symbols (i.e., words, numbers, and pictures) and create new ideas, like database solutions, that are most valued in an information economy. It is only through our language that information can be created, conveyed, and stored. You've heard of information overload. Well, we've gotten very efficient at creating and sharing new ideas, in large part due to computers and the Internet.

Those of you preparing for careers in technical occupations such as engineer or scientist will be writing a lot because your jobs will entail manipulating language (including mathematic symbols) to generate new ideas. Simply put, much of your daily activity will be mediated through written communication. One study of engineers and scientists who worked at a plastics company demonstrated that the technical employees at all levels of the organization—from staff, to supervisors, to upper management—spent 40 to 50 percent of their weekly labor engaged in writing-related activities, including drafting, revising, and editing.[1] The staff engineers and scientists spent a significant portion of their writing-related time *drafting* reports about their technical work. The upper mangers, on the other hand, spent most of their time *editing* the proposals and reports written by staff and supervisors. All levels of engineers and scientists at this company also spent a significant portion of their time in other writing-related activities including reading and responding to e-mails and letters, planning and recording meetings, and preparing oral presentations.

If you want to be successful, you had better be a good writer. If you desire to eventually be promoted to upper management, you not only need to be able to write but to critique and improve the writing of others.

Effective Technical Writing Is Rhetorical

The best way to improve your writing is to carefully and thoroughly understand your writing situation, including who your readers are. Techniques for analyzing your writing situation come from the field of rhetoric. Achieving good writing, as you read above, takes more than paying attention to correct spelling and grammar. This course assumes, by the way, that you already possess a certain level of basic literacy that includes knowledge of spelling and grammar. (You've presumably taken both high school and freshman level English classes.) Certainly, grammar and mechanics are important for effective communication. Too many basic mistakes and your readers may have trouble understanding your message or, worse, they may stop trying to understand it. However, even the most error-free prose could fail to convey technical information if it doesn't take into consideration the following:

- What is the purpose for writing?
- Who is the reader?
- What is the context?
- Who is the writer?

What Is the Purpose?

Technical communication occurs in social contexts and therefore is a social action. People write technical documents because they want to do things in the world: the social action involved in technical writing connects the intangible sphere of thought with the concrete realm of human activity. A person authoring online help for a word processing program, for example, aims to help users understand a specific computer program and, in turn, write more effectively. Likewise, an engineer writing a 40-page recommendation report detailing a six-month-long study of groundwater patterns near a proposed factory informs crucial decisions about the factory's construction—decisions that will impact multiple individuals and groups such as the local community where the proposed factory will be built, the global company looking for U.S. operations, and the engineering firm conducting the study. If a technical document did not intend to initiate, effect, or affect action, then it probably wouldn't be written. Because technical communication is social, it is often understood differently by different people for different reasons. Technical writing often fails because writers have ignored the social nature of technical communication.

All technical writing has multiple purposes, and the more the writer is aware of the multiple purposes of a given writing task the better the writer can evaluate the document's effectiveness. These purposes include the following:

- To inform
- To describe
- To persuade
- To educate

- To summarize
- To warn
- To indemnify
- To document/create a record
- To maintain goodwill

The most common purpose of technical writing is to convey information. Scientific research articles present new knowledge based on original research. Engineering specifications describe a product in precise detail so it can be used, reproduced, repaired, manufactured, or tested. Instruction manuals inform users of the procedure for assembling a product and how to use it safely, including warning of any hazards associated with improper use of the product. There are other purposes of technical writing. Manuals also train or educate users about the best ways to accomplish a task using the technology documented in the manual. (Our culture, however, is very suspicious of manuals because of their history of being so badly written. Ever hear the expression, "When all else fails, consult the manual?") In this class you'll learn how to analyze your writing situation by first asking, "What is my purpose for writing?"

It is helpful to also be aware of secondary purposes. Think of product warnings. A warning's primary purpose is to inform the reader of a danger or hazard associated with misuse of the product. However, in addition to informing the user of the hazard, the warning must educate the user about how to avoid the hazard. Moreover, the warning serves to indemnify the company from product liability, protecting the company from litigation if a user gets injured using the product. Thus, it is in the best interest of the company to write a warning with sufficient information so that it persuades the reader to use the product properly. Oftentimes companies are afraid to include too much information in warnings, particularly about the potential for serious injury, for fear it will scare customers away. However, it is safer—for the company and the users—to be up front and honest about the risks associated with using a product. (Can you identify the multiple purposes of product warnings discussed in this paragraph?)

Another way to think about a document's purpose is to consider how the reader will use the information. Think of the bad software documentation example used previously. If the writer had imagined how the reader would be using the information, i.e., sitting in front of the computer and wondering how to get the new program working as soon as possible, he or she might have written the instruction more directly, in a way that the reader could understand.

You should also always consider as one of your purposes the following: to maintain goodwill, or to keep positive relations with the reader. Poorly written technical documents alienate readers from the technology and disincline them to the makers of the technology. Whose fault is it that people have trouble setting the clock on a VCR or using all the features of a cell phone, the instruction writer's or the user's? A lot of technical writing is about relations: who will do what job, who will fund which project, who will give you the time you need to finish a task, who will finish the test you need to confirm your product modifications? Engineers and scientists interact with all types of people day in and day out. Technical professionals must treat readers humanely, with respect and dignity, regardless of the reader's technical background and level of specialized knowledge.

Who Is the Audience?

The biggest factor to consider when relating technical information is the audience's level of knowledge or expertise about the information (see Figure 1.1). Are your readers experts who have the same background and familiarity with your subject as you, or are they a laypeople who are unfamiliar with the technology you are writing about and who do not have the same technical training as you? The less your audience knows about the subject, the less technical your document should be. The less your readers know about your subject, the more you will have to explain and the more you should avoid highly specialized language. It's impossible to avoid all technical language, so you may have to define certain terms.

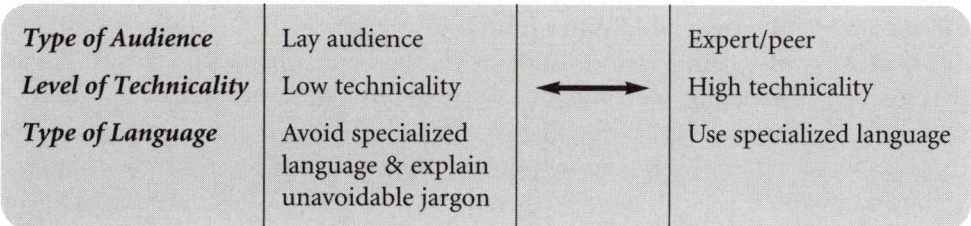

Figure 1.1. *Adjust the Language of the Message to the Type of Audience*

It is also helpful to judge your audience's receptivity to your message. You might be writing a progress report about some product testing that is behind schedule and over budget. How will you convince your skeptical boss to allow you more time and money on the project? Some audiences are predisposed one way or the other to your message. For instance, writers of environmental impact statements have to contend with both supporters and opponents to proposed land-use projects. Anticipating the kinds of questions that readers will ask of your documents (e.g., How is your costly product testing going to save us money in the long run? How will the landfill affect the nearby housing development?) will help you adjust your message accordingly. If your audience is a group of people, or if you don't know your immediate audience, sometimes you have to make educated guesses about the audience's general needs and interests.

In today's global economy it is also important to take into consideration cultural factors that may affect an audience's ability to understand your message. Will your document be read by an international audience? Do you need to tailor your document to accommodate a worldwide audience?

Finally, there are always primary and secondary audiences to a document. Your boss may be the primary audience of a report you write, but your colleagues may also read the report. You might also be thinking of your professional peers as an audience if you wish to reuse some of the information in the report for a journal article you hope to publish. You must also always consider lawyers, the public (people outside the company you work for), and the media as audiences. Several scientists at the controversial Yucca Mountain Nuclear Waste Repository were found to have admitted falsifying important quality assurance records in e-mails related to research on the safety of storing nuclear waste 100 miles away from the city of Las Vegas, Nevada. Whether these scientists' actual research is valid is being debated. However, legislators, the public, and the media learned of these e-mails, calling into question the integrity of the scientists conducting these important safety studies.

What Is the Context?

Context means considering any factors related to your writing situation—the setting you are writing in—that might affect your document. For instance, you should consider when the document is due and what form the document should take. Should it be written as a lab report, specification, manual, tutorial, memo, or letter? Given the type of document called for, how should it be organized and formatted? Each research field, for instance, has different standards for citing previous research and they tend to differ from one another. Each company has slightly different ways of writing various documents. Many companies keep style guides that dictate document specifications such as margins, headings, and proper spelling and abbreviations. While this course will teach you some basic guidelines for good technical writing, it cannot possibly anticipate and teach all the variations among companies. You can often learn formats of company-specific documents by collecting samples of previously written documents.

When considering your writing context you should also consider more abstract cultural, ethical, and political factors related to your situation. There are more women than ever in technical fields. Technical communication must therefore be written using nonsexist language. The audiences of technical communication are more diverse and international than ever. Technical writing also often deals with subjects that have consequences for real human beings. Medical professionals must keep accurate patient histories. Airline technicians must keep accurate airplane maintenance records. Engineers must determine the safety of everyday products. Scientists must not falsify experimental procedures to achieve desired results. Writers of technical documents are often confronted with difficult ethical decisions that have financial, emotional, and physical health implications for the writer, the writer's organization, and the public. Lastly, consider if there are any political factors that could affect the document. Will you be stepping on anyone's toes? How will you report a costly error to your boss? How will your message impact others, positively or negatively? Are there any financial or legal factors that you should consider?

Who is the Writer?

This sounds like an odd question to ask yourself before you write, but this question is designed to get you to think about your relationship as a writer (and a person) to your reader. The writer of the confusing accounting software manual forgot that he or she was a computer scientist writing to a layperson. He did not consider that the word *populate* means something different to non-computer scientists. Consider what relationship you have, or want to have, with your audience and then consider how you should adjust your message:

- Who are you writing to, your boss, your employees, the public, your profession, your teacher?
- Who are you writing as, a technical specialist, a manager, an employee, or a student?
- What tone (or mood/personality) should you use to convey your message?
- What language is acceptable for you to employ as a writer?

In general, your tone should always be balanced and professional, never hostile or arrogant. But should your writing be more formal, avoiding first-person "I" and second-person "you," or can it be more informal, addressing the reader directly? Most memos and letters are written using first and second person. It is also becoming more acceptable to use first person in technical reports (e.g., instead of writing, "The reaction of silicon dioxide films were studied," write, "We studied the reaction of silicon dioxide films.") When considering how you will communicate your message as a writer you can also think about any specific terms that will need to be translated or avoided.

Conclusion

In the previously mentioned study of technical writing at a plastics factory the researchers found that managers used the engineers' and scientists' writings to evaluate overall thinking ability, not just writing skills. "Their writing reflects the quality of my staff's minds," one supervisor commented. "Management wants original ideas…and weak or confused documents simply call attention to incompetence."[2] Since most of the work of a technical job is conveyed and documented in proposals, reports, and other memoranda, it should be no surprise that your writing will play a crucial role in how others judge you as a coworker and employee. In this sense, you should always aspire to produce good technical writing, writing that accomplishes its purpose, is sensitive to the needs and background of its readers, shows awareness of the context of the document, and is written in a manner that causes readers to perceive the writer as thorough, professional, and humane.

Exercises

1. In class, teams of 4 or 5 create a simple Lego object and write instructions on how to assemble it. Exchange the disassembled Lego's and instructions to see which teams can produce the desired object. Discuss what made teams' instructions effective or ineffective.

2. Go to Darren Barefoot's "Hall of Technical Documentation Weirdness" online at http://www.darrenbarefoot.com/hall/main.php. Individually or in groups, browse the examples. Find an example and sketch and/or write a clearer version of the chosen example.

3. Individually or in a group, find a current or past event in the news that relates to technical writing/communication. Use the Internet and your library's periodical indexes. Reflect on how the news story relates to issues discussed in this chapter.

4. Interview someone you know who writes in the workplace. Ideally, interview someone in the field you plan to enter. Brainstorm a list of interview questions in class such as what types of documents are written, how much time is spent writing (including planning, drafting, and revising), and the type of training the writer received (in school and at work). Write a profile of that workplace writer based on your interview and share it with the class.

End Notes

[1] James Paradis, David Dobrin, and Richard Miller. "Writing at Exxon ITD: Notes on the Writing Environment of an R&D Organization" in *Writing in Nonacademic Settings*, ed. Lee Odell and Dixie Goswami, 285. (New York: The Guilford Press, 1985).

[2] Paradis et al., pp. 295–296.

Chapter 3
Professional Writing Style

Style in writing refers to **rhetorical form** rather than grammatical form. That is, style is a matter of choosing words and constructing sentences in a way that suits the writer's and, more importantly, the readers' tastes. These tastes have developed over time and are more about social custom and habit than any abstract set of grammatical principles.

Effective business and technical writing style should be clear and concise. Because busy business readers—managers, workers, clients, and consumers—don't have the time and patience to deal with long, hard-to-read prose you want your writing style to be direct, readable, and easily understandable. You want readers to comprehend your points without having to waste time deciphering confusing sentence structure, word choice, or grammar.

Communicating effectively begins foremost with having a firm understanding of your *purpose* (what you are trying to achieve with your writing) and your *audience* (who you are communicating with). Once you clarify these for yourself, as a way to measure if you have communicated your purpose given a particular audience's knowledge and ideals, you can work on organizing your message, writing a rough draft, determining how well it accomplishes its purpose, then editing it for clarity and concision.

Revision versus Editing

Remember the difference between **revision** and **editing**. Revision happens when you make changes to the higher-order issues of purpose, organization, and development of details/content. Writing effectively means first getting your initial plans on paper, then revising them to improve their organization and development based on the reader's needs and anticipated questions.

Editing, on the other hand, represents lower-order changes to style, grammar, punctuation, and spelling. When first putting your thoughts to paper or computer screen don't worry about lower order concerns. Focus on getting your thoughts on paper in whatever form they come to you. Write out your ideas as if you were explaining them orally to a close, respected friend in a casual setting. Novice writers get hung up on crafting perfect prose, word for word, sentence by sentence. Rarely should you have to stop a thought in midsentence to search for the perfect word. Put down whatever word first comes to your mind and then, during the revision and editing stages, go back and change it if necessary. Worry first about the big picture, about getting your point out of your head and down in print, then go back and massage the text for clarity and concision.

Principles of Effective Professional Writing

After revising your document—be it a letter, memo, case analysis, or report—for purpose and audience, edit your prose carefully to conform to the following principles of effective professional writing style:

1. Use plain, conversational language (write clearly)
2. Use active voice, generally
3. Eliminate unnecessary words (write concisely)
4. Guide readers through your documents
5. Use direct presentation, generally
6. Use parallel grammatical structure
7. Keep your tone polite and professional
8. Write like a human being
9. Be diplomatic
10. Pay attention to usage, spelling, grammar, punctuation, names, and numbers

Use Plain, Conversational Language (Write Clearly)

Writing clearly means organizing your message and choosing your words in a way that optimizes the readers' chances of easily understanding your message. To write clearly: (1) keep it simple; and (2) keep it conversational. You're not writing to show off or impress, but to inform and persuade. Lack of clarity, variously called gobbledygook, affectation, or double-speak, has crept into our way of communicating for many reasons, including the fear that plain messages don't sound important enough or won't impress subordinates or superiors. Gobbledygook, a term coined in 1944 by Congressional Representative Maury Maverick, refers to impenetrable corporate or government prose that is used to variously cover up a poorly conceived message, needlessly impress, deliver unpleasant information or, as in the case of jargon, to alienate or confuse laypeople (see Figure 3.1).

> **Instead of:**
>
> To ensure that the new system being developed, or the existing system being modified, will provide users with the timely, accurate, and complete information they require to properly perform their functions and responsibilities, it is necessary to assure that the new or modified system will cover all necessary aspects of the present automated or manual systems being replaced. To gain this assurance, it is essential that documentation be made of the entities of the present systems which will be modified or eliminated.
>
> **Write:**
>
> Make sure to document all changes to the current system so any mistakes can be corrected. (1)

Figure 3.1. *To fix gobbledygook, say what you mean and get to the point*

The easiest way to write clearly (unambiguously, in a way that is easy for the reader to understand) is to write as you talk. You want the *voice* of your writing to sound professional, yet not artificial. It is acceptable to use first person "I" and second person "you" in business memos and letters and some reports. You can also use contractions (two words joined by an apostrophe, e.g., *can't* and *don't*) in all but the most formal of correspondence (e.g., when applying for a job). Write as you talk and your writing will significantly improve, but avoid slang, which is unprofessional.

Cut these signs of gobbledygook from your writing:

- **Overly inflated or pompous word choices**. Instead of *effectuate*, write *do*. Instead of *fabricate*, write *make*. Instead of *initiate*, write *start*.
- **Clichés**. Clichés are those trite, wordy expressions used so often that we don't even know what they mean. Instead of *as per your request*, write *I have completed* or *Here is the information you requested*. Instead of *pursuant to*, write *about* or *regarding*.
- **Jargon with lay audiences**. Jargon is: (1) confused, unintelligible language; strange, outlandish, or barbarous language; (2) technical terminology or characteristic idiom of a special activity or group; (3) obscure and often pretentious language marked by circumlocutions and long words.
- **Buzzwords**. Buzzwords are trendy business terms that slip into an industry or profession, such as *robust*, *metrics*, and *synergy*.

Use Active voice, Generally

Verbs are either *active* (The manager *assigned* the task) or *passive* (The task *was assigned* by the manager). Active verbs immediately follow the subject and make clear who or what the center of action in the sentence is: "*The manager* assigned..." Passive sentences use verb forms of *to be*, including *is, am, are, was,* and *were*. Sentences written in active voice are usually more concise and easier to understand than sentences written in passive voice.

Computerized grammar checkers can identify passive voice constructions and ask you to revise them to a more active construction. There is nothing inherently wrong with the passive voice, but if you can say the same thing in the active mode, do so.

> For more help, particularly advice about when passive constructions are preferable, see the Purdue Online Writing Lab's "Active and Passive Verbs" handout, which can be found at:
> http://owl.english.purdue.edu/handouts/grammar/g_actpass.html.

Eliminate Unnecessary Words (Write Concisely)

Writing concisely is a way to achieve clarity by making your point in the fewest possible words. It means eliminating unnecessary and redundant words. This is often referred to as eliminating clutter, deadwood, or flabbiness. Novice writers tend to inflate their prose with extra adjectives, adverbs, and phrases. Efficient, uncluttered writing is more easily understood and appreciated because it doesn't waste the reader's time. Words worth cutting include:

- Passive constructions, verb phrases that include a form of *to be*, such as *am, is, was, were, are,* or *been*
    - Instead of *The report was distributed by the committee*, write *The committee issued the report*
- Most adjectives and adverbs
    - Instead of *Rogers convincingly explained his position*, write *Rogers explained his position*
    - Instead of *positively certain*, write *certain*
- Weak modifiers
    - Cut words like *very, rather,* or *little*
    - Instead of writing *you can do a little better*, write *you can do better*
- Meaningless words
    - Cut words like *kind of, actually, really,* and *basically*
- One part of a doubled phrase
    - ~~full and~~ complete, first ~~and foremost~~, ~~any and~~ all, each ~~and every~~
- Redundant phrases
    - ~~period of~~ time, ~~basic~~ fundamentals, ~~past~~ history, ~~future~~ plans, ~~free~~ gift, ~~terrible~~ tragedy

- Compound prepositions
  - Instead of *at this point in time*, write *now*.
  - Instead of *despite the fact that*, write *although*.
- Obvious words
  - When it arrived, I cashed your check immediately
  - Our company president, who ultimately makes all of the decisions, should be notified about this.
- Wordy clichés
  - Use *because* instead of *due to the fact that*
  - Use *if* instead of *in the event that*
  - Use *must* instead of *it is incumbent upon*
  - Use *that you requested* instead of *as per your request*

A good habit to get into is this: after you have revised your draft to a point where you think you're done, **try to eliminate at least 20% of its total word count**. Does this sound excessive? It is the equivalent of editing a 10-word sentence down to 8 words. This percentage is just a rule of thumb. As you become more adept at eliminating useless words from your writing the percentage you cut may increase. The point is that good writers always pay attention to eliminating clutter. It is a service paid to readers.

Guide Readers through Your Documents

One of the best ways to improve comprehension is to add graphic highlighting and other subtle page design elements to make your documents more readable and scannable (faster to read through). Novice writers see these elements all the time but often fail to use them to improve their own documents. Add elements such as the following:

- **Summaries**, or overviews at the beginning of documents and sections within documents
- **Transitions**, words that help the reader see connections between ideas, such as *however*, *another*, *next*, and *thirdly*
- **Letters** such as (a), (b), and (c) within text
- **Numerals** such as 1, 2, and 3 listed vertically
- **Bullets**, raised periods, black squares, or other figures
- **Headings and subheadings** signaling the topic or main idea of particular sections
- **Emphasis** by creating contrast through **boldfacing**, *italicizing*, or underscoring can highlight main ideas or important points. You can also alter the size of text or use ALL CAPS. Be careful not to overuse emphasis.
- **Smaller paragraphs**, as readers don't like to struggle through dense, packed prose. Break up larger-sized paragraphs. It's okay to have one, two, or three sentence paragraphs.

Use Direct Presentation, Generally

Busy readers want to quickly determine whether they should read something now or later, if at all. Let readers know up front what they're reading. State and summarize your main points before developing them. Always include an opening or overview section in your documents that states your point in the first sentence or two. Don't wait until the end of the opening, as you were taught in writing introductions to academic essays. Headings and bulleted points can help present arguments directly, e.g., "Because of the recent corporate restructuring our firm must adopt the following new policies: 1) ... 2) ... 3) ...." Another example of directness in business writing is the type of organization called the *managerial style*, which presents recommendations at the beginning of a report. The name of this style derives from the decision-making manager's desire for clarity up front.

However, being direct isn't always the best tactic. In some cases, as when you have to deliver bad news or persuade a skeptical reader, a more indirect approach usually works better. In bad news letters this is called the buffer, a letter opening that begins with a positive but relevant statement before introducing the the bad news. Another instance where an indirect approach would be necessary would be in reports that deliver information to skeptical or hostile readers, such as a recommendation report calling for major increases in expenditures. In this case it would be preferable to first make your case and then present the recommendation toward the end of the report.

Use Parallel Grammatical Structure

Express similar ideas in balanced or parallel constructions. Writers often string together a number of items, activities, etc. into one sentence. For example, a writer might list goals for a project: "By quarter's end, we should increase sales, eliminate overhead, and cut losses." This sentence is parallel because each goal is stated using the **same verb form**, i.e., *increase, eliminate,* and *cut.* A sentence is unparallel when it fails to present similar items in the same grammatical fashion.

You can use this principle to present several points within a paragraph as a bulleted or numbered list:

The position is available to applicants that meet the following criteria:

- *Business-related degree, preferably a BS/MBA*
- *5–6 years working in computer-related industry*
- *High degree of experience interfacing directly (face-to-face) with customers and middle-level management*

Keep Your Tone Polite and Professional

The principle aim of all professional communication is to establish and maintain relationships. Without the goodwill of your audience you cannot motivate, persuade, sell, etc.—all that you're trying to accomplish in business. Avoid language choices that are angry, condescending, arrogant, or rude. Avoid words with negative connotations such as *cannot, forbid, fail, prohibit,* and *deny.* Instead of *poor service,* write *service,* as in, "I'm writing to ask your help in addressing the recent service I received from your agency." This often requires a shift in perspective. Instead of complaining, why not set out to inform your reader about the situation and politely ask for a remedy? This is often called taking a *you* approach. Avoid sounding self-centered; make the reader the main focus. Instead of, "This change addresses our labor costs associated with sloppy renters," write, "This change enables us to clean vacated properties faster, thereby insuring prompt service to you the customer."

Write Like a Human Being

As language expert William Zinsser writes, "Just because people work for an institution they don't have to write like one."1 Your writing will be better if it treats readers with respect and attempts to connect with them as humans. The words you choose should project an image of yourself as respectful, thoughtful, warm, and personal. Your writing should focus on the needs and concerns of the reader. This is called the ***you* approach**. Show empathy for the readers' point of view, provide reasons for whatever it is you're asking them to do, and be polite and friendly. Instead of writing, "A sales receipt must accompany any refund," write, "To reduce the processing time of your refund, please provide a copy of the sales receipt." If you have to deliver bad news or compel the readers to act in ways to which they may react negatively, provide reasons why the bad news or unfavorable action is necessary.

Be Diplomatic

Use positive words to express negative ideas. Instead of writing, "I want to complain about your poor customer service," write, "I wish to bring a customer service matter to your attention." Avoid words that suggest the reader is dishonest, careless, or mentally deficient. Be indirect in situations where you must deal with sensitive topics, persuade skeptical readers, or deliver bad news. Lastly, avoid putting anything in writing you might regret later.

Pay Attention to Usage, Spelling, Grammar, Punctuation, Names, and Numbers

Documents should always be proofread carefully. Remember, proofing a document should come at the very end, after you've sufficiently stated and developed your point and improved the style of its expression. The rule of thumb for checking surface-level issues like *usage* (using the right word—it's vs. its; affect vs. effect; principle vs. principal) or punctuation is this: if you're not 100% certain of the correct usage or rule, consult a reference guide or someone who knows the rule. Expert writers rely on reference aids as a basic tool when writing, whether to double-check a tricky rule, look up an unfamiliar word, or answer questions like, "Should I capitalize this word here?" With many reference tools online (e.g., Bartleby.com) this task has become faster.

To proofread more effectively:
- Allow adequate time to proofread. Build time for proofing into your schedule.

- Give yourself a break from the document. Come back to it after a period of rest, such as after lunch or the following day.
- Print a copy and read through it at least twice. Try reading backwards, bottom-to-top from the last page.
- Read the document out loud to yourself. You'll often catch constructions that read okay but don't sound right.
- For high-stakes documents read the draft to someone or have them read the draft back to you. It is commonplace for expert writers to give and receive feedback, especially in the workplace.
- Use, but *do not trust*, computerized spelling and grammar checkers. Have you read Jerrold Zar's "Ode to a Spell Checker"?

    For the rest of the limerick, go to http://www.jokes2go.com/poems/9892.html?

Draft, Revise, Take a Break…Then Revise Again

Experienced writers draft and revise rather than attempt to write a document all in one sitting. Few writers can craft perfect prose word for word. Initial drafts often contain what's known as **writer-based prose.** This is writing that satisfies the writer but not the reader. It contains:

- Long, rambling sentences that make sense only to the writer
- Lengthy paragraphs covering too many topics
- Sequential items in hard-to-read paragraph form rather than easier-to-read bulleted lists
- Ideas presented in a narrative, chronological structure rather than direct or indirect organization patterns

This kind of writing must be revised into **reader-based prose**, or writing that "foregrounds and makes explicit the information a reader needs or expects to find."[1] The more you can visualize an actual conversation with your reader, the more you can anticipate the kind of questions the reader might need answered, the kind of information the reader needs to know, and the best order to present that information to the reader.

Since you must revise to achieve effective reader-based business writing, here are some strategies for revising your initial draft:

1. **Test your draft against your initial plans**: Did all of your goals and good points from your notes make it into the draft? Did you forget any key ideas? Did any irrelevant ideas creep into the draft?
2. **Use clues that reveal your plan to your reader**: Did you include enough direct and explicit statements about your plans to your reader (e.g., opening purpose statement)? Did you include other cues like headings and transitions to help clarify your point to the reader?

3. **Keep the promises you made in your writing**: Did you follow through on any promises you made to readers. For instance, if your overview states that you will provide four recommendations each supported by an example, did you keep that promise?
4. **Shift your perspective to that of the reader**: According to one guide written by Purdue's Online Writing Lab, to revise effectively you must distance yourself from the draft and see it objectively from the readers' perspective. "Unless you divorce yourself from the paper," states the guide, "you will probably remain under its spell: that is, you will see only what you think is on the page instead of what is actually there." To see your writing objectively you need to build time into your writing process so that you can set your draft aside and return to it later, either after a break or the following day. The guide includes a detailed checklist for helping you assess your drafts from a reader's point of view. The following table summarizes that checklist and can be used to evaluate any business writing, from short memos or letters to longer reports:

Table 3.1.

	Reader-Based Revision Checklist[2]
Detail	Have I included all the information the reader needs to know, and nothing more?
Organization	Have I arranged topics in a way the reader expects the topics arranged, either directly or indirectly depending on the reader's favorable or negative disposition to my message?
	Have I compartmentalized the topics of my message?
	Have I included headings and transitions between topics to help guide my reader through the presentation of my message?
Language	Have I used the right words to get my message across, or are some words unnecessary, elevated, or in need of a definition?
Tone	Have I chosen words that present me to the reader as courteous, respectful, rational, professional, and ethical?
	Have I avoided negative language and used positive words to express negative ideas?
	Have I focused on the reader's needs and interests rather than my own (the you approach)?
Mechanics	Are there any glaring typos or mechanical mistakes that would diminish my credibility as a writer and businessperson in the eyes of the reader?
	Have I carefully proofread and corrected all spelling, grammar, and punctuation mistakes?

Exercises

1. Revise the following sentences for clarity and concision. Use the word counts in parentheses as a guide; see if you and a partner can revise the sentence to at or below the suggested word count *without altering the meaning of the original.*

   a. There are many words that are useless that can be eliminated through revision that is carefully done. (17 words to 5 words)

   b. After extensive exhaustive labors on the assigned project task, Roger needed to take a long respite. (16 to 10)

   c. In the event that customers who have questions about their automobiles call us, we have just created this really useful program that is intended to save time in our customer service responses. (33 to 7)

   d. The report was written by engineers out of the taskforce from Atlanta.

   e. Until such time as we are in possession of more information, we will be unable to offer a satisfactory reply to your inquiry. (12 to 6)

   f. We are unable to provide you with access to computerized laboratories in view of the fact that you are not in possession of personal identification. (26 to 8)

   g. Until such time as our company changes this policy, we are not in a position to supply information for the purpose of aiding your investigation of other companies. (29 to 11)

   h. In the event that employees fail to utilize the requisite safety precautions, our optimum course is to transmit this information to OSHA and request directives before deleterious effects ensue. (30 to 13)

   i. People renting our apartment units are concerned about whether we would provide for replacements of damaged personal property in the event of accidental discharge of the sprinkling system in the apartment complex. (33 to 15)

   j. Subsequent to my conversation with Mr. Jones via telephone, the physician at the hospital where Mr. Jones is hospitalized indicated that Mr. Jones has an enlarged kidney and must undergo immediate surgery if a cure is to be effectuated. (40 to 18)

End Notes

1. Flower, Linda, and John Ackerman. "Evaluating and Testing as You Revise." *Strategies for Business and Technical Writing* (4th ed). Kevin J. Harty. New York: Allyn and Bacon, 1999.

2. "Revision in Business Writing." Purdue Online Writing Lab. URL: http://owl.english.purdue.edu/handouts/print/pw/p_revisebus.html

Chapter 4

Business Correspondence

The principles for writing letters and memos are very similar. The audience and occasion determines whether a letter, memo, or e-mail is necessary. The advent of e-mail has blurred the traditional separation of memos, which have been for internal communication, and letters, which have been for formal, external correspondence.

Memos

Memos are generally for *internal correspondence* within a company (see Figure 4.1). When you need to communicate in writing to someone within your own company, you'll write a memo. But memos often extend beyond the boundaries of a company to include correspondence with customers and clients, especially as e-mail blurs the boundaries between memos and letters.

	Memo	**Letter**	**E-mail**
Audience	Internal	External	Both
Formality	Informal	Formal	Informal
Social Cordialities (salutation, complimentary close, signature)	No	Yes	Yes

Figure 4.1. *Memo, Letter, E-mail Comparison: This table lists the generic tendencies for each form of correspondence, but all three forms are more or less interchangeable depending on the situation and organization.*

Because they are primarily intended for in-house audiences, memos typically dispense with the social cordialities of business letters such as using a formal salutation ("Dear Mrs. Doe"). But that does not mean memos are any less important. Memos inform people of decisions, policies, procedures, problems, solutions, responsibilities, and deadlines—all while creating a record of a company's activity.

Print and electronic memos often replace oral meetings in instances where face-to-face communication is impractical or undesirable. While most memos range from one to three pages long, reports written in memo format can be dozens of pages long. For a sample memo, see Figure 4.2.

Annotations	Memo
In heading, use complete names and position titles	**To:** All employees, SeaCorp Inc. **CC:** Roger May, VP Operations **FR:** Elaine Darling, Chief Executive Officer *ED* **DT:** January 4, 2004 **RE: E-Mail Use Policy**
Use a short, specific subject line	
Add an opening that states purpose clearly	The following policy covers appropriate use of any e-mail sent from a SeaCorp e-mail address and applies to all employees, vendors, and agents operating on behalf of SeaCorp.

Rationale

Our company needs to implement e-mail use guidelines for the following reasons:

1. **Professionalism**: by using proper e-mail language, our company will convey a more professional image and avoid tarnishing the company's public image.

2. **Efficiency**: e-mails that get to the point are much more effective than poorly worded e-mails.

3. **Protection from liability**: employee awareness of e-mail risks will protect our company from costly lawsuits.

E-mail Etiquette

It is far too easy to treat e-mail communications as an informal manner of communicating; however, it is a written record that is maintained in the ordinary course of business. You are reminded to maintain a business-like and professional decorum to your e-mail correspondence:

- Format e-mails as you would print memos
- Use proper spelling, grammar, and punctuation.
- Never send messages containing derogatory, defamatory, obscene, or inappropriate content or attachments

A simple test to bear in mind is: if you would not write the e-mail content in a formal business letter, then refrain from the use of such content in your e-mail messages.

Company Monitoring

All electronic mail messages are the property of SeaCorp, and employees should have no expectation of privacy whenever they store, send, or receive electronic mail using the company's electronic mail system. E-mail is a business record and may be subject to review by your manager, other employees, the courts, governmental agencies, litigants and other persons who are not the intended recipients of the e-mail. Deletion of a file on an employee's computer does not delete a record of the e-mail from the company's system.

Personal Use

Using a reasonable amount of SeaCorp resources for personal e-mails is acceptable, but non-work related e-mail must be saved in a separate folder from work related email. Sending chain letters or joke e-mails from a SeaCorp e-mail account is prohibited. Virus or other malware warnings and mass mailings from SeaCorp shall be approved by SeaCorp's VP of Operations before distribution. These restrictions also apply to the forwarding of mail received by a SeaCorp employee.

Enforcement

Any employee found to have violated this policy may be subject to disciplinary action, up to and including termination of employment.

Questions about this policy should be directed to the VP of Operations.

Annotations (continued):
- In body, break up big "chunks" of text into smaller, more readable paragraphs
- Use headings, lists, and emphasis (bold) to increase readability
- Notations can appear in either the heading (e.g., the "CC" line) or at the end of a memo
- To sign a memo, either type or handwrite your name or initials after the closing. You can also initial by your name in the heading (as this sample does)
- Add a clear closing

Figure 4.2. *Sample Memo*

Letters

Letters are generally for *external correspondence* between a company and its clients, customers, or other interested parties. Letters are generally used when a more formal document is needed, such as legal proceedings, promotions, terminations, and thank yous. Thus, letters can also be written to members inside an organization when the occasion calls for it.

As noted in Figure 4.1, letters include the traditional social conventions of the salutation, complimentary close, and signature. These elements lend are what lend a formal air to the letter. (See Figure 4.3 for a sample letter.) These elements can also be used to make a memo or e-mail seem more formal. Electronic mail is being used more and more for external correspondence, but it is still considered bad form to use e-mail at times when very formal correspondence is needed. Don't quit or fire anyone in an e-mail!

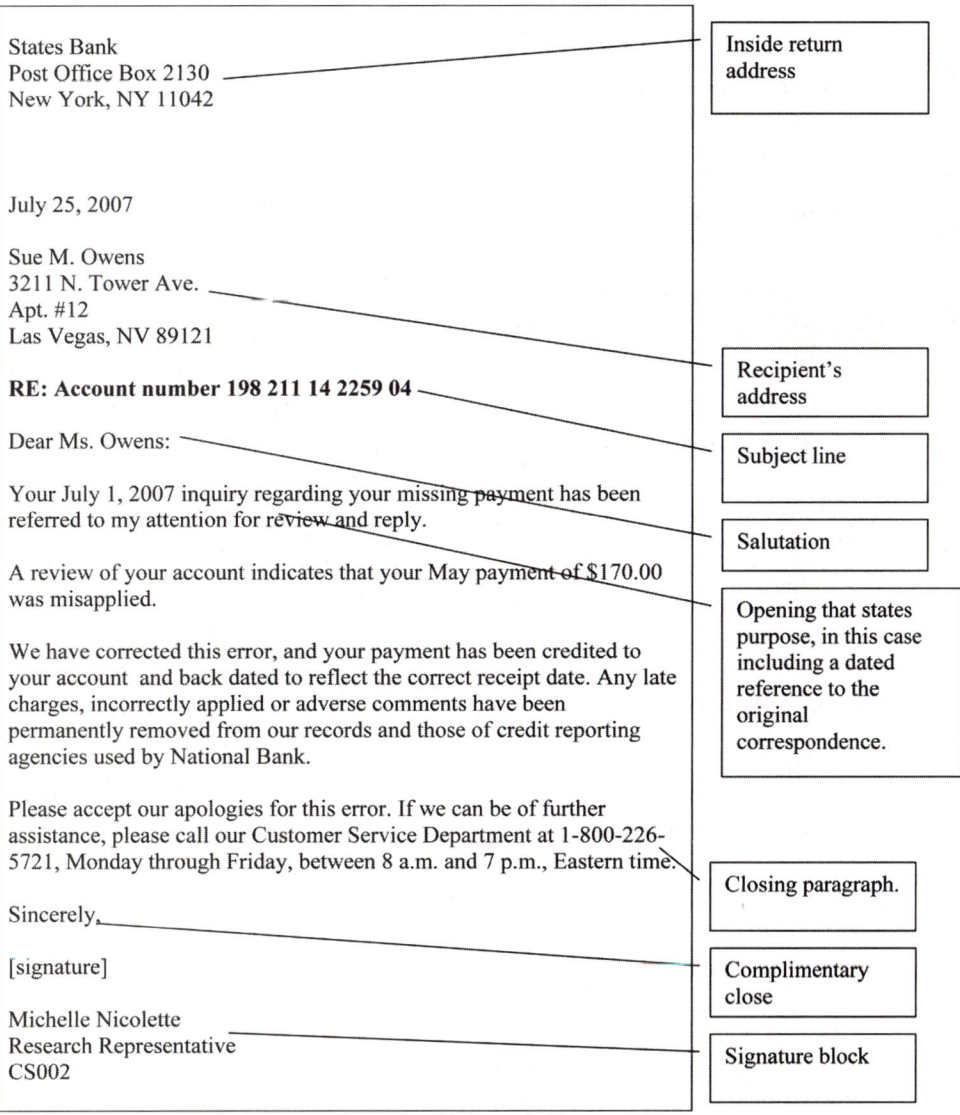

Figure 4.3. *Labeled Diagram*

Chapter 4: *Business Correspondence*

E-mail

Perhaps one of the biggest changes the computer has brought to business is e-mail. Electronic mail was invented in 1971 but didn't take hold as a ubiquitous communication tool until the late 1980s, when its use exploded. One source estimates that the average employee sends and receives over 170 messages daily, and that number should double by 2010.[1] Another study estimated that business managers send and receive twice as many e-mails as front-line employees.[1]

The problem with e-mail is that it is rapidly replacing all other forms of communication, including phone and face-to-face conversations. Add in personal e-mails and spam, and employers are spending too much time sifting through new e-mails and hunting for important information in old e-mails. Many experts are writing about a general backlash against e-mail.[5]

E-mail is best for simple requests, statements, questions, or acknowledgements. (See Figure 4.4). It is also good for sending large documents as attachments.

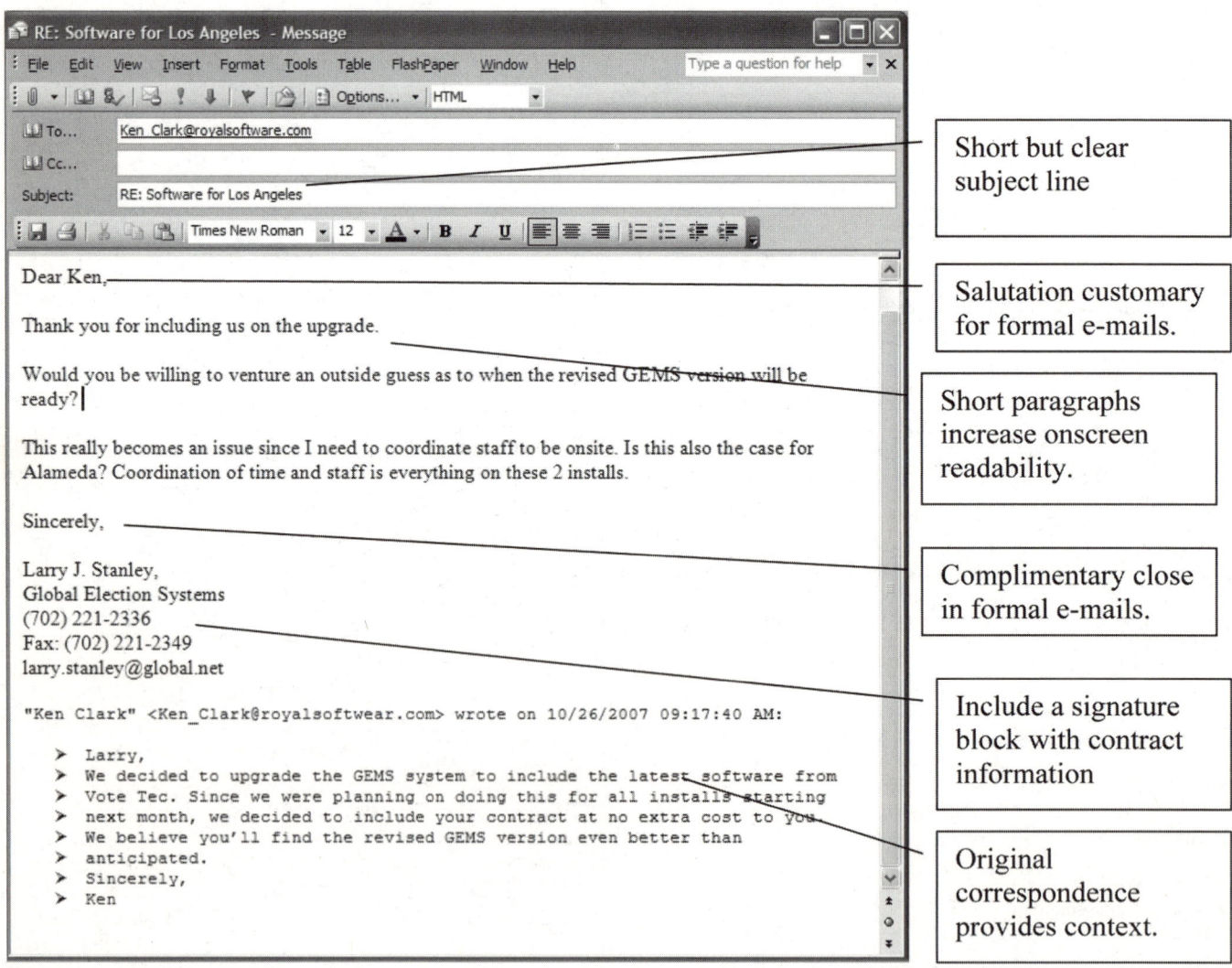

Figure 4.4. *Sample e-mail*

E-mail is not good for conveying mass amounts of information, subtle emotion, humor, or sarcasm. Discussing long and complex issues via e-mail can actually delay decision making. It is better to have a phone conversation or meeting and use e-mail to confirm points discussed in person.

E-mail is particularly poor at conveying subtle emotions. Hence, e-mail should not be used for sensitive situations such as major announcements, firings, job evaluations, or negative criticism. Too many people treat e-mail as an informal communication medium similar to speech, and use business e-mail to send all kinds of unprofessional messages including tirades, whinny complaints, and personal attacks on co-workers.

Generic Structure of Memos, Letters, and E-Mail

The basic structure, or organization, of letters and memos is based on how business readers approach such documents. When planning your letter or memo, always imagine a busy reader—busy boss, busy coworker, busy customer, or busy client. Readers don't have time to pore over your whole document to figure out why you're writing and what you want from them. The basic structure of a memo or letter includes the heading, opening, body, and closing. Any memo or letter you write should use this structure to as quickly and clearly as possible address basic questions readers have when reading any correspondence (see Figure 4.5).

Memo or Letter Parts	Answer Reader Questions
Heading and Opening:	Who is this from?
	What is this about?
Body:	What do I need to know about this?
	Why is this important to me?
Closing:	What do you want me to do, and when?

Figure 4.5. *Basic Structure of the Reader-Centered Memo or Letter*

The Heading

The first parts of the correspondence answer the reader's question, "What's this about?" The heading contains the particulars about the sender and receiver of the document. It also includes information about when it was written and what it is about. Heading formats vary from memo to letter, and from organization to organization, but most generally contain some or all of the following information:

- Complete and correctly spelled names
- Proper job titles

- Correct date
- Business addresses
- Short but informative subject line

You can compare Figures 4.2 and 4.3 to see the difference between the letter and memo headings. The letter heading includes the sender's and recipient's complete mailing addresses. The memo includes the "to" and "from" lines. Both letter and memo headings include a date and should include a subject line. Note that business mailing addresses include job titles and company names, in addition to the person and street address.

The reader uses the information in the heading to determine when, if ever, the memo or letter should be read. Business readers prioritize what to read based on the sender and subject lines. Blank, vague, or clichéd subject lines might cause confusion or be ignored. While you can add details such as "IMPORTANT" and "URGENT" to a subject line to catch a reader's attention, you should avoid doing this to every letter, memo, or e-mail you send.

The Opening

The opening or overview section of a memo or letter is like the introduction to an essay. After scanning the heading, the reader looks to the first few paragraphs to further determine the document's purpose and whether it deserves the reader's full attention immediately, later, or ever. When writing the opening, be sure to:

- State the purpose for writing in the first sentence or two (Note: You don't have to write "The purpose is…", just state the purpose, e.g., "This memo presents plans for restructuring . . .")
- Summarize the main points of the memo
- If you're responding to a request or inquiry, reference the date and subject (e.g., Here is the report you asked for last Thursday, April 28…")
- If you're addressing a problem, include details about the problem
- If you're making recommendations or requesting readers take action, include details of the recommendations or request

The Body

The body or discussion section of the memo or letter develops the information noted in the opening. The body answers the reader's questions, "What do I need to know?" and "Why is this important?"

There are two basic ways to organize information in the body of a memo or letter:

1. **Direct organization**: This arrangement puts the most important points first, followed by supporting information. For instance, if you're writing a recommendation report to address a problem, first provide the recommendations and then discuss any research that you conducted to address the problem. (This plan assumes you briefly summarized the problem in the opening of the memo or letter. You might need to include a more

detailed discussion of the problem in the body, after the recommendations section.)

2. **Indirect organization**: This arrangement puts supporting details first, leading to your main point. If you expect readers to react skeptically or negatively to your main points, put them at the end of the memo. This way, you can use supporting details like the statement of the problem and discussion of alternatives to prepare readers for your conclusion.

Make sure, as Figures 4.1 and 4.2 show, to use the following devices to help guide your readers through your discussion and make your memo easier to read:

- **Use shorter, single-spaced paragraphs**. Break up big chunks of text into smaller, more readable paragraphs. It is okay to have paragraphs that are one, two, or three sentences long. Memo paragraphs are usually single-spaced with a double-space between paragraphs. Paragraphs may or may not be indented, depending on a given company's format.
- **Compartmentalize** each different point or topic in a separate paragraph. As you revise your memo, if you find a paragraph that has more than one main point or is too long, consider splitting it into two or more paragraphs.
- **Use section headings** that indicate the main point of each section, helping readers scan the contents and organization of the document. You can use headings in documents of any length, and you should definitely use them in memos and letters longer than half a page.
- **Use bulleted or numbered lists** to present itemized or sequential topics, such as multiple recommendations, the steps in a procedure, or a series of questions. Avoid overusing bullets, however. A separate bullet for every paragraph of your memo can make it difficult to distinguish important points.
- **Use text emphasis** such as bold or italics to highlight key words or important information. Use text emphasis sparingly, however. Highlighting too much text in a document defeats the purpose of using emphasis because nothing stands out anymore.

The Closing

The closing reiterates for the reader key points discussed in the correspondence answering the questions, "What do you want me to do?" and "When do you want it done?" Since most business writing seeks some kind of action from the reader, make sure you state this action clearly in the closing. Include specific information about when you want this action accomplished. Avoid vague, clichéd endings such as "If you have any questions, please feel free to contact me." If you want a response, then state so unequivocally: "Please respond to my request by noon on Friday, April 28." Be polite and courteous in your closing, especially when asking people to do things for you or on your behalf.

Letters include a signature block that includes a complimentary close, ink signature, and signature line (see Figure 4.3). Memos do not have to have a signature block at the end because the writer's name appears in the heading. However, many memos include the writer's initials, first name, or full name either typed or handwritten at the end of the memo or next to the "From" line in the heading (see Figure 4.2). Always include a signature block in electronic correspondence (see Figure 4.4).

Notations

Memos and letters often contain additional information in either the heading or after the closing:

- **Copies**: "CC" (*carbon copy*) indicates other recipients of the memo. Include full names and position titles for all you copy. This way, others can see if, for instance, an attorney or senior executive received a copy. "BCC" (*blind carbon copy*) means those who received CC copies won't see who received the BCC.
- **Enclosures**: Notations such as *Enclosure, Encl.,* or *Encl.: Report (15 p.)* indicate and create a record of any additional documents included with the memo. It's best to include specific information about the enclosures such as title and number of pages. It's bad form to write "Attachment."
- **Initials**: Initials at the end of a document such as "GK/lm" indicate the writer (in all capital letters) and the typist (lowercase letters).
- **Multiple pages**: If your memo is longer than one page, include page numbers on every page after the first page. Other information such as the date of the correspondence, the company name, the subject of the memo, or the name of the writer is often added, which helps create a professional appearing document.

Correspondence Style

As discussed in Chapter 3, Professional Writing Style your business writing should be conversational, clear and concise, human, and diplomatic.

Conversational

It's acceptable to use first person "I" and second person "you" in business memos and letters. You can also use contractions (two words joined by an apostrophe, e.g., *can't* or *don't*) in all but the most formal of correspondence. Write as you talk and your writing will significantly improve, but avoid slang, which is unprofessional.

Clear and Concise

To write clearly (unambiguously, in a way that is easy for the reader to understand), avoid letting gobbledygook slip into your writing. *Gobbledygook* is the term of art for overly pompous, affected language that is long-winded and difficult to understand. Instead use plain, everyday language. Rather than writing *utilization* write *use*. Depending on your audience's familiarity with technical terms, also limit your use of jargon and buzzwords.

A good habit to get into is this: after you have revised your draft to a point where you think you're done, try to **eliminate at least 20% of the total word count**. This percentage is just a rule of thumb. As you become more adept at eliminating useless words from your writing, the percentage you cut may increase. The point is that good writers always pay attention to eliminating clutter. It's a service paid to readers.

Human

As language expert William Zinsser writes, "Just because people work for an institution they don't have to write like one."[2] Your writing will be better if it treats readers with respect and attempts to connect with them as humans. The words you choose should project an image of yourself as respectful, thoughtful, warm, and personal. Your writing should focus on the needs and concerns of the reader. This is called **the *you* approach**. Show empathy for the reader's point of view, provide reasons for whatever it is you're asking them to do, be polite and friendly. Instead of writing, "A sales receipt must accompany any refund," write, "To reduce the processing time of your refund, please provide a copy of the sales receipt."

Diplomatic

Being direct isn't always the best tactic. Be especially tactful and polite when conveying bad news or attempting to persuade skeptical readers. Use positive words to express negative ideas. Instead of writing, "I want to complain about your poor customer service," write, "I wish to bring a customer service matter to your attention." Avoid words that suggest the reader is dishonest, careless, or mentally deficient. Lastly, avoid putting anything in writing you'd regret later.

Common Genres of Correspondence

Many memos and letters are written for occasions common enough that they can be classified into the following categories:

- Inquires
- Claims and adjustments
- Rejections and acceptances
- Transmittals and cover letters

Inquires

Common reasons for writing inquires are to find more information about something from the recipient. You may want to learn more about a product or service, so you write the company requesting more information. You may want to know if a company can offer some service beyond what it ordinarily provides, such as helping a school or charitable organization with a donation of money, goods, or services. You may even want to know if a company has any unadvertised job openings. All of these are occasions that prompt an inquiry memo, letter, or e-mail.

When writing an inquiry, follow these guidelines:

1. State your reason for writing in the opening of the letter.
2. Identify who you are and why you need the requested information.
3. Use a numbered or bulleted list to outline specific information you desire or questions you have.
4. If possible, offer to compensate the recipient for replying, e.g., by paying phone or mailing costs, to acknowledge the recipient in your work, or to provide the recipient with a copy of whatever you are working on, such as a report, charity program, etc.
5. If persuasion is necessary, provide a tactfully worded reason how the recipient will benefit by providing the information you desire.
6. In the closing, thank the recipient for their cooperation.

Claims and Adjustments

Common reasons for writing claims are to request some compensation that you are due or inform a recipient of some problem and attempt to persuade the recipient to provide some remedy. If you were not happy with a stay at a hotel, and weren't satisfied with how the front desk manager responded, you might write to the corporate headquarters requesting compensation. The "adjustment" is the response by the recipient that informs the claim-writer of how the claim is being resolved. Claims are sometimes also referred to as complaints, but not all situations that require a claim are necessarily occasions for complaint. Furthermore, as the Business Writing chapter indicates, the act of complaining has negative connotations and often puts recipients on the defensive. Instead of angrily "complaining about poor service," the goal should be to calmly "bring a customer service matter to the company's attention."

Claims almost always involve argumentation and subtle persuasion, since you are usually asking for a remedy that goes beyond what the recipient wants to provide. It is important you build a case for your claim by providing an accurate summary of the problem, reasons why you believe compensation should be provided, and a clear statement of the compensation you desire. Avoid overly negative, hostile, or insulting language. Most individuals want to be fair and provide quality service; however, people are less inclined to help irate customers and are even willing to lose the irrational customer. As the saying goes, you can catch more flies with honey than with vinegar.

When writing a claim, follow these guidelines:

1. State the reason you are writing, in generally positive language, e.g., to request compensation or bring a customer service matter to their attention.
2. Provide a detailed description of the situation or problem. Be specific about dates, incidents, model or part numbers, etc. (you may also which to provide documentation as evidence to further support your claim, e.g., copies of receipts, contracts, correspondence, photos).

3. Provide reasons why your request should be granted. You can appeal to the recipient's sense of fairness, commitment to quality, or desire for customer satisfaction.
4. State what specific compensation you desire, making sure the compensation is in proportion to the claim.
5. Close politely.

When writing the adjustment, or reply to the claim, follow these guidelines:

1. Reference the date of the original claim letter and state whether you are granting or denying the claim in the opening.
2. Acknowledge the claim-writer's concerns and efforts for writing.
3. If you deny the request, explain the reasons why the request can not be granted.
4. If you deny the request, consider offering some partial or alternative compensation.
5. Conclude politely, reiterating you or your company's values regarding company.
6. Maintain a professional and tactful tone throughout the letter. Avoid sounding defensive, defiant, arrogant, or resentful.

Rejections and Acceptances

Common occasions for rejection letters are having to inform someone of bad news such as not being hired in a job search, being denied a promotion or raise, or being laid off from a company. An adjustment letter denying a claim is another form of rejection. The same rules of tactfulness, professionalism, and sincerity apply. Most people can handle rejection or bad news if they are treated like a human being. Providing sincere reasons often lessens the sting of rejection.

When writing a rejection letter, follow these guidelines:

- Be thankful and appreciative in the opening.
- Provide a tactfully worded rejection in the opening or second paragraph.
- Provide specific reasons for the rejection, if possible.
- End politely, trying to leave the door open for a future relationship.
- Avoid using generic or canned language. Part of treating people human is not resorting to stereotypical or clichéd language. Be as concrete and sincere as possible when providing reasons for rejection or bad news. Be tactful and indirect, but don't "sugarcoat," as most readers can see through false sentiment.

Acceptance or good news letters are easier to write and often only require a few short words of congratulations. Recipients of good news letters usually shift gears to quickly wondering about the next step, such as when the job starts, if the company will pay for relocation, how the compensation will be delivered, etc. Try to anticipate such questions and provide suitable answers. Good news recipients also appreciate being given reasons for their good fortune. Giving details about the number of applicants reviewed or number or entries in a competition heighten the sense of accomplishment and make it possible to use the letter as documentation of achievements in future settings, such as a career portfolio.

Transmittals and Cover Letters

Transmittal correspondence informs the recipient of delivery of something else, usually a document such as report or a resume. More information about writing transmittals can be found in Chapter 25, The Client Project.

A cover letter is a specific form of transmittal letter used in the job search to accompany the resume. More information about writing cover letters can be found in Chapter 7, Cover Letters.

Writing Electronic Correspondence

Remember that e-mail is a permanent written record that is generally regarded as company property. Companies are becoming increasingly wary of how much time and money is wasted on poorly written or inappropriately used e-mails. To write more effective e-mails, follow these guidelines:

- **Use a clear subject line** that makes it easier for the reader to understand why they received the e-mail and that make it easier to search in the future. It is particularly important to avoid vague subject lines in e-mails. Take a keywords approach that strings the key words of the message into a meaningful phrase.

- **Use all of the devices for readability** previously discussed for memos and letters. Use shorter paragraphs, lists, and text emphasis to make the e-mail easier to read.

- **Format your e-mails similar to a letter**. Formal e-mails should include a salutation line such as, "Dear Ms. Smith:" and a signature block at the end that includes a complimentary closing, your name, title, and contact information.

- **Consider when a different communication medium would be preferable**. Many companies are encouraging employees to use the phone more to conduct business that is too complex to be dealt with efficiently in back-and-forth e-mail exchanges. Use the phone when a personal touch is needed, the message is too nuanced to be conveyed in emotionless e-mail, or the information is too sensitive to be recorded in an e-mail. Many companies also encourage the use of Instant Messaging (IM) chat for activities such as meeting planning.

- **Keep the length of e-mails short**. Knowing that employees and managers receive dozens of e-mails daily should remind you to keep your e-mails short. Most e-mails should not be longer than one screen. Busy readers do not have the time or inclination to read dense e-mail messages. Some companies use e-mail to circulate denser reports, but longer documents should be attached to the e-mail.

- **State the reason you are writing in the beginning**, just as you should with memos and letters. Don't bury key details in the middle of paragraphs and at the end of the message.
- **Remember the difference between personal e-mail and business e-mail.** Business e-mails should be treated as formal as print memos, in terms of style, language, and mechanics. Though it depends on how well you know your audience, you should generally avoid slang and IM speak (the abbreviations, acronyms, and phonetic spellings used in chat and text-messaging) and instead use proper grammar and punctuation.
- **Be professional**. Partly because e-mail is used for personal reasons too, people confusingly think of e-mail as a more conversational and less permanent medium than print memos and letters; however, it is just the opposite. E-mails are easier to retrieve than most print documents, and they are potentially saved longer! Only write what you would feel comfortable saying to the person's face. Don't send confidential or sensitive e-mails that you don't want people other than the recipient to read. Don't write anything nasty about coworkers or bosses, and don't send venting e-mails that complain, whine, or reprimand.
- **Check for misspellings, poor grammar, and typos**. Be sure to edit your business e-mails, particularly when the message needs to be more formal and is being sent to someone who doesn't know you.
- **Double-check before hitting "send."** For very important e-mails you may even want to print the draft for closer editing. Most e-mail applications allow you to save drafts. You could save the draft, print it out for closer reading, and return to it after a break so you can re-see it from a fresh perspective.
- **Be familiar with your organization's e-mail use policy**. The courts have been very clear that your company e-mails are not your private property. Most companies have e-mail use policies that notify employees of x, y, and z. Most companies archive all e-mails (even when they are deleted off of your work PC) and many companies monitor company e-mail for inappropriate content.
- **Don't let e-mail run your (work) life!** Turn off the auto-alert function so you don't find yourself checking e-mails constantly. Block out specific times of the day to check e-mail, such as in the morning, after lunch, and before leaving the office. Don't get in the habit of replying to e-mails immediately. Decide if e-mail is the best way to communicate and don't copy the world on all your messages.

Exercises

1. Revise the memo below:

   Acme Inc. Memorandum

   To: All Division Managers

   From: Stanley Rogers, Vice President, Admin.

   RE: Policies

   CC: Thomas Rickley, Manager, Building Services

   The garage on the south side is being renovated to include a new third level and 150 new parking spaces. Because of these renovations, a new parking policy will be in effect between the months of February and December of this year. Please circulate this memo to all of your division employees and make an announcement at your next Division meeting. Between February to July, Administration and Finance employees can continue to use the first-level of the south side garage. Textile and Packaging employees can continue to use the second-level of the south side garage between Feb. and July. Service and Manufacturing employees can use either the second level of the south side garage or park in the south side lot at this time. (It is recommended that first- and second-shift Manufacturing employees use the south side lot.) The south side garage will be closed to all employees between August and December. At that time, Administration must use the front lot. Finance employees must use the east side lot. Service and Manufacturing employees must use the east side lot, and Textile and Packaging employees must use the west side lot. Employees with special circumstances or handicaps should contact Thomas Rickley, Building Services Manager (ext. #7765), for a front lot permit during the renovations. All other questions should be directed to the Building Services office.

2. Find a sample business memo or letter from your work or personal life. (Make sure it does not contain any confidential or proprietary information.) Write an analysis of the document's rhetorical situation (context, writer, purpose, audience) and how well written the document seems to be given the rhetorical situation. Identify if the memo or letter is written in a common genre and include a critique of the document's format and style in your analysis. Consider whether the format and style contribute to the overall effectiveness of the document.

3. Individually or in small groups, copy the contents of the sample documents from exercise 2 into Word and run a word count. Then revise the memo to make it more concise. Aim to eliminate at least 20% of

> the word count without affecting the overall intended meaning of the memo. This may require eliminating some unnecessary words, rewriting excessively long sentences, and using other devices, such as shorter paragraphs, lists, and headings. OPTION: Choose one memo in particular and have the entire class work on and compare the same example.

End Notes

[1] Harvey, Michael. "Concision." *The Nuts and Bolts of College Writing.* http://www.nutsandboltsguide.com/concision.html

[2] Zinsser, William. "Writing in Your Job." *Strategies for Business and Technical Writing* (4th ed). Kevin J. Harty. New York: Allyn and Bacon, 1999. p. 70.

[3] Flower, Linda, and John Ackerman. "Evaluating and Testing as You Revise." *Strategies for Business and Technical Writing* (4th ed). Kevin J. Harty. New York: Allyn and Bacon, 1999.

[4] "Revision in Business Writing." Purdue Online Writing Lab. URL: http://owl.english.purdue.edu/handouts/print/pw/p_revisebus.html

[5] Brady, Diane. "*!#?@ the E-Mail. Can We Talk?" *Business Week.* New York: Dec 4, 2006. 109.

[6] Petrillo, Alan M. "Make Your E-mail Matter." *Office Solutions.* July 2007. 30-32.

[7] Lamb, Sandra E. "Office E-Mails and Voice Messaging." *Business and Economic Review.* Oct.-Dec. 2006. 30-31.

Chapter 5
Reports

The following provides an overview of reports as a type of professional writing. It includes a discussion of report organization and design, including page layout, style sheets, and graphics.

Reports as a Type of Professional Writing

All reports address a particular need by conveying in-depth information, often aimed at persuading readers to act on that information. Reports are written for many purposes, including:

- **Proposals**. Proposals written by private companies for government contracts can be hundreds of pages long, aimed at persuading the government agency the company is capable of providing the desired service in a cost-effective manner. Such proposals include detailed discussion of the company's qualifications, feasibility of the work, necessary methods and materials, production schedules, and budgets.
- **Business plans**. A type of proposal, business plans document the scope of a business venture and persuade investors of its feasibility. Such plans include a description of the proposed product or service, technical explanations of technologies used, market analyses, production requirements, facilities and personnel needed, projected revenues, funding requirements, legal issues, qualifications of start-up personnel, and investment potentials.
- **Feasibility and recommendation reports**. These reports provide in-depth analysis helping readers to make informed decisions. Feasibility reports consider if a proposed action is practically, economically, and technologically possible. Recommendation reports compare two or more options against similar criteria and advocate for a particular option.
- **Primary research reports**. These reports detail the results and implications of research conducted in experiments, surveys, and other field work. Research and development companies produce primary research reports as the first step in considering practical applications of advances in science and technology.

- **Technical-background reports.** Often called *white papers*, these reports provide background on a given topic that readers need to understand better, such as new technologies, market projections, or other business trends. Technical-background reports typically argue a particular position or solution, but in an objective, fact-based manner.
- **Usability reports.** A combination of primary research and recommendation reports, usability reports detail the results of user testing conducted to ascertain the user-friendliness of a prototype or working Web site.

A key feature of reports is that they are written for different kinds of readers and **different types of reading**. Studies show, for example, that executive decision makers spend most of their time reading the executive summary, the detailed synopsis appearing at the beginning of the report. Decision makers may not ever read the entire report. However, some may decide to look more closely at parts of the report. For example, a manager may want a closer look at the recommendations, so after reading this executive summary she will scan the table of contents and then skip to the recommendations section. Others read the whole report more closely. Someone might be charged with carefully studying the main points of the report and formulating an opinion about whether to follow the report's recommended action. In this case, the person may read more closely from front to back, but he or she may also decide to skip from section to section. In fact, most readers do not read reports linearly, or from front to back. Instead, they skip around from the opening to the conclusion to the appendices, etc. Many of the **purposefully redundant** elements are designed to facilitate these different ways of reading complex persuasive documents (e.g., the report is summarized more or less in several places: the transmittal letter, the executive summary, the opening, and conclusion). Each report section also includes clear summaries that inform the reader of the purpose of that particular section.

Each section of the report contributes to its overall persuasiveness. A report's background persuades the reader that there is indeed a problem or need that calls for action. The report methodology establishes that the writers have carefully thought about how best to study that problem. The findings and recommendation sections aim to convince readers that the writers carefully collected and analyzed information in light of the problem or need being addressed. Appendices help document careful research and provide details that ordinarily would disrupt the flow of the main report discussion.

> You can review the following links to get a sense of the variation among report writing practices:
>
> - Market Feasibility Study
>   http://www.ndsuresearchpark.com/pdf/ndsu_study.pdf
> - California Energy Commission Reports
>   http://www.energy.ca.gov/reports/
> - Securities Industry Recommendation Report
>   http://www.nasd.com/web/groups/rules_regs/documents/rules_regs/nasdw_006434.pdf
> - Marketing Research White Paper
>   http://publisher.yahoo.com/rss/RSS_whitePaper1004.pdf
> - OneStart Portal Usability Report
>   http://www.indiana.edu/~usable/reports/test3_report.pdf
> - Research Reports from the National Commission on Writing
>   http://www.writingcommission.org/report.html

Report Format

Reports vary in length and format. Most reports are the result of considerable investment of time and resources, so careful attention is paid to dressing up the report in the trappings of commercial publishing. That is, the reports will include elements that one typically finds in published books, such as a cover, table of contents, and pleasing page design. Some reports, however, are distributed in memo format and therefore dispense with more formal elements like a cover. Except for memo (or e-mail) reports, most reports contain the following elements:

Front Matter	Transmittal Letter
	Cover (or Title Page)
	Executive Abstract
	Table of Contents
	List of Figures
Body	Overview/Introduction
	Background
	Recommendations
	Method
	Findings
	Closing
Back Matter	References
	Appendices

Front Matter

Front matter includes the transmittal letter, cover, executive summary, table of contents, and list of figures.

- **Transmittal Letter.** The letter of transmittal is a cover letter that accompanies the delivery of documents to external audiences. The letter should follow standard letter format and be limited to one page. The letter should state the purpose of the report, highlight pertinent recommendations, and briefly describe research methods. Include a courteous closing that expresses willingness to work further with the reader and offers to answer any questions. This letter can either be bound with the report (inside the cover) or delivered with the report as a separate document.
- **Cover.** The cover (or title page) includes the title of the report, date, names of the people writing the report, and name of the recipient. It might include the address of the client company. If it is a report for school, the cover could also include your instructor's name and course information. To add to the report's overall persuasiveness some attention should be paid to the cover's visual appearance. You can find sample report cover and page designs on the Internet.
- **Executive Summary.** After reading the executive summary, the reader should know the gist of the whole report, including background, methods, findings, and recommendations. This condensed version of the final report summary is for decision-making audiences who lack the time to read the entire report closely. It should be written in lay terms, designed for managers who may not have the technical expertise of other readers. Because the reader is most interested in what your final recommendations are, be sure to emphasize this part of your executive summary with subheadings and/or lists. You may wish to omit a summary of the findings, mentioning relevant findings only as you present each recommendation in some detail. The executive summary should not exceed 2 pages and can be placed either in the front matter (before the table of contents, paginated with lower-case Roman numerals) or in the body of the report (as page 1).
- **Table of Contents.** The table of contents provides readers with an overview of the organization of the report and a way to find the information that they want quickly and efficiently. List all major headings and all subheadings. Use double spacing, indentation to indicate subheadings, and leader dots. Technical reports often include a decimal numbering system for easy reference (e.g., 1.0, 1.1, 1.1.1, 1.1.2, etc.)
- **List of Figures.** The list of figures is a separate page after the table of contents that lists the titles of all visuals with page references. Tables are listed separately from figures. Consult models and style guides for specific ways to format the table of contents and list of figures.

Body

The body, or main part of the report includes the following sections: introduction, background, recommendation, methods, findings, and closing. This arrangement, which presents the recommendations early in the report, is known as **managerial**

organization because it emphasizes suggestions for action over methods and data. The managerial organization is recommended for most reports other than primary research reports, which should follow the traditional scientific introduction, methods, results, discussion arrangement.

- **Introduction**. Review the gist of the report, focusing on the statement of the problem, pertinent background/history of the project, results, and recommendations. Include a summary of the report's major sections.
- **Background**. You should include the background information from your Design Plan, including a description of the report's purposes and audiences, your preliminary evaluation, and testing goals. The background section should be formatted with a first-level (L1) heading, while the preliminary evaluation and testing goals sections should be subsections formatted with second-level (L2) headings. In usability reports, you may want to add a screenshot of your site's home page in the background section.
- **Recommendations**. Make several recommendations to improve the issue based on the trends you noted in the data. Be sure to support your recommendations with references to your data (e.g., "As the results indicated...," or "Because of this, we recommend...."). Provide suggestions on how to fix the problems you identify. Arrange this section with specific subheadings that summarize each recommendation (e.g., "Recommendation #1: Sell Unnecessary Assets"). Consider adding visuals to help clarify recommendations. One simple visual is a figure that lists the recommendations.
- **Method**. Describe whatever research methods were used to address the question or problem motivating the report. Primary research reports provide a narration of the research procedures used.
- **Findings**. This section presents the results of any primary research conducted.
- **Conclusion**. End the report with a persuasive appeal that suggests the benefits of accepting the report's claims. If applicable, discuss any limitations of the current study or anything that could use further analysis or research.

Back Matter

References and appendices appear at the end of the report.

- **References**. If you paraphrase or quote any outside sources to support your work, you must document these sources with in-text citations and a list of references. If you have fewer than 3 citations, cover them fully in a parenthetical reference in the text or with a footnote. If you have more than 3 sources, use a list of references. Be sure all references follow an accepted style manual. Chicago style and APA are the most common. (See the style guide section below for links to information on documenting sources.)
- **Appendices**. Appendices are for supplemental information that ordinarily would interrupt the flow of the main discussion of the report. Be sure to reference appendices in your discussion. For example, when you

describe your test methods include a reference to your data sheets (e.g., "see Appendix A: User Test Protocol") and include a copy as an appendix. Avoid putting graphics in appendices, since they typically help clarify key points. For each appendix make a divider page with the appendix title or put the appendix title at the top of the first page of the appendix.

Report Design

Reports pay careful attention to visual presentation. Aside from aiding readability, careful presentation demonstrates the investment of organizational resources into a document that will be read by many different people within and outside an organization.

Page Numbering

Paginating reports follows traditional publishing practice:

- All pages are numbered except for the front and back covers and the transmittal letter. Page numbers are not always displayed.
- All front matter pages are numbered with lowercase Roman numerals (e.g., i, ii, iii, iv).
- The report body begins on page 1 and is labeled with Arabic numerals (e.g., 1, 2, 3, 4).
- The executive summary is either included in the front matter before the table of contents (labeled with lower-case Roman numerals) or as the first section of the body (labeled as page 1 with Arabic numerals).
- Page numbers are not displayed on the transmittal letter, cover, first page of the table of contents, page 1 of the introduction, and appendix divider pages (if used).
- You can place page numbers either centered or to the right of either top or bottom margins. You can also include the report title and, if applicable, a company name in headers or footers.

Page Layout

When multiple writers contribute to a document the best way to ensure consistency is to use a style sheet. Style sheets dictate font styles, heading formats, and page layout (see Figure 5.1). You can download a Word file of a sample style sheet for use with your final report, or you can follow similar guidelines, examine some sample reports, and design your own.

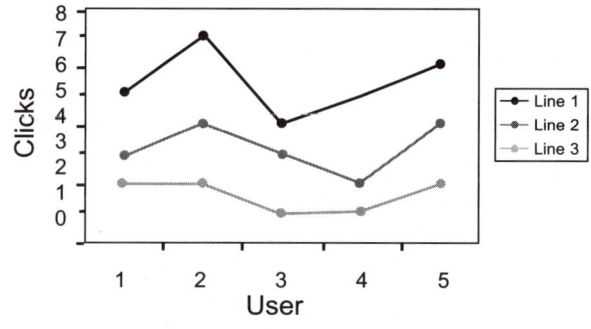

Figure 5.1. Sample Style Sheet. This version uses 1.5" left margins for L1 headings and 2" left margins for body text. This creates visually pleasing white space and allows for binding

Graphics

Whenever you want to emphasize a point, catch your reader's attention, or clarify a complex idea, add a visual graphic. Next to section headings, captions on graphics are the second most read parts of a report. This underscores the importance of adding visuals to important documents like reports. The following discusses types of visuals and conventions for integrating visuals within the prose discussion of the report.

Tables

Tables are good for showing precise data, comparison, and juxtaposition of abstract ideas. The most common type of table presents data in rows and columns like an Excel spreadsheet. Another form of table is called a matrix, which presents text in tabular form. For example, Figure 5.2 is a matrix table comparing several computers according to various specifications.

Table 1. Best Value Desktop PCs. *Dell offers the best all-around value. For a few more dollars, Gateway offers more speed and storage. Sony and iBuyPower: too pricey.*						
Computer	Processor	Hard Drive	RAM	DVD	Price	Comment
Dell Dimension 4600	2.8 GHz Pentium 4 CPU	120GB hard drive	128MB NVidin GeForce FX 5200 graphics	16X DVD-ROM and 48X CD-RW drives	$1129	Best price, comes with an LCD screen, good speakers. Performance is good.
Gateway 510XL	3-GHz Pentium 4 CPU	160GB hard drive	128MB NVidin GeForce FX 5200 G graphics	8X DVD±RW and 16X DVD-ROM drives, media-card reader	$1400	Big hard drive, fast processor. Gateway warranty is a hidden extra cost
Sony VAIO PCV-RS520	3-GHz Pentium 4 CPU	160GB hard drive	128MB TI Radeon 9200 graphics	8X DVD±RW and 16X DVD-ROM drives, media-card reader	$1529	Very nice design, excellent performance. Keyboard is sub-par and sound system is weak
iBuyPower Back To School	2.2 GHz Athlon 64 3400+ CPU	80GB hard drive	256MB ATI Radeon 9800 XT graphics	8X DVD±RW drive, media-card reader	$1694	Weakest performance and hard disk. Best RAM, but cost is high.

Source: PCWorld.com. Top 15 Desktop PCs. August 2004.

Figure 5.2. **Sample Matrix Table.** *Some squares have precise data, but other squares include words, allowing for easy comparison of different criteria. Notice that captions are located above table graphics*

Charts

Some distinguish between *graphs*, which show values plotted on an x and y coordinate system, and *charts*, which do not; but the terms are often used interchangeably. Charts include the following:

- **Bar graphs**—show comparison and contrast.

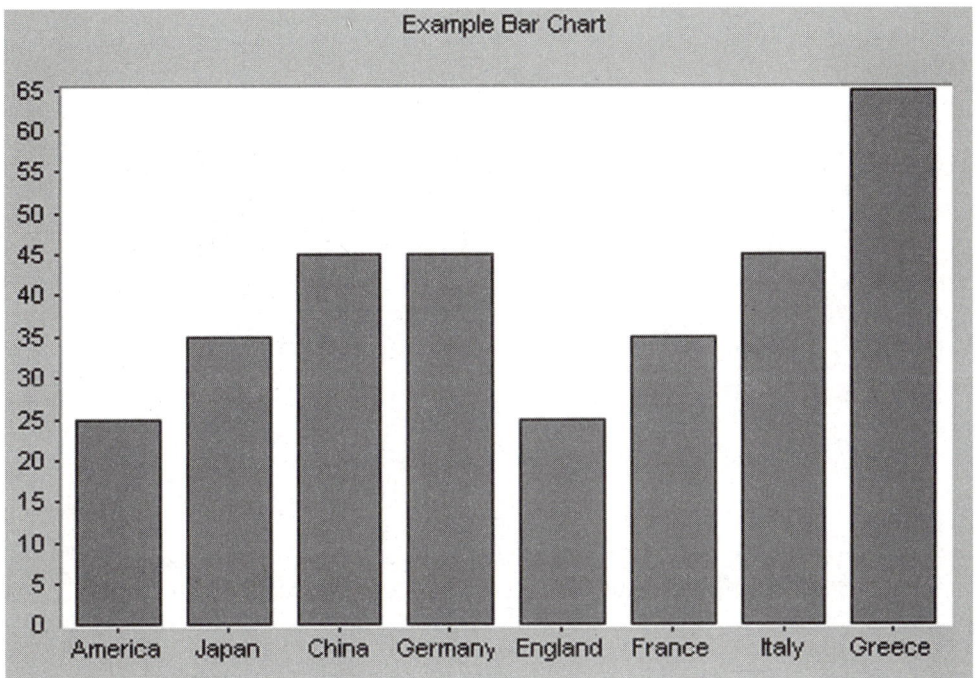

- **Line graphs**—show trends.

Chapter 5: *Reports*   59

- **Pie charts**—show percentages, or parts of a whole.

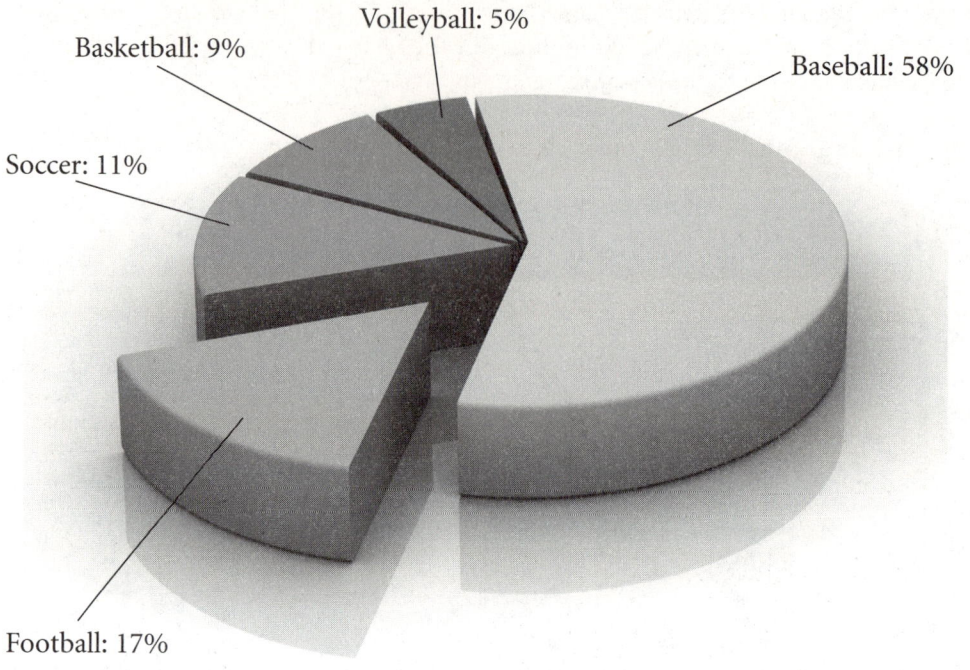

- **Organizational charts**—show structure or hierarchy.

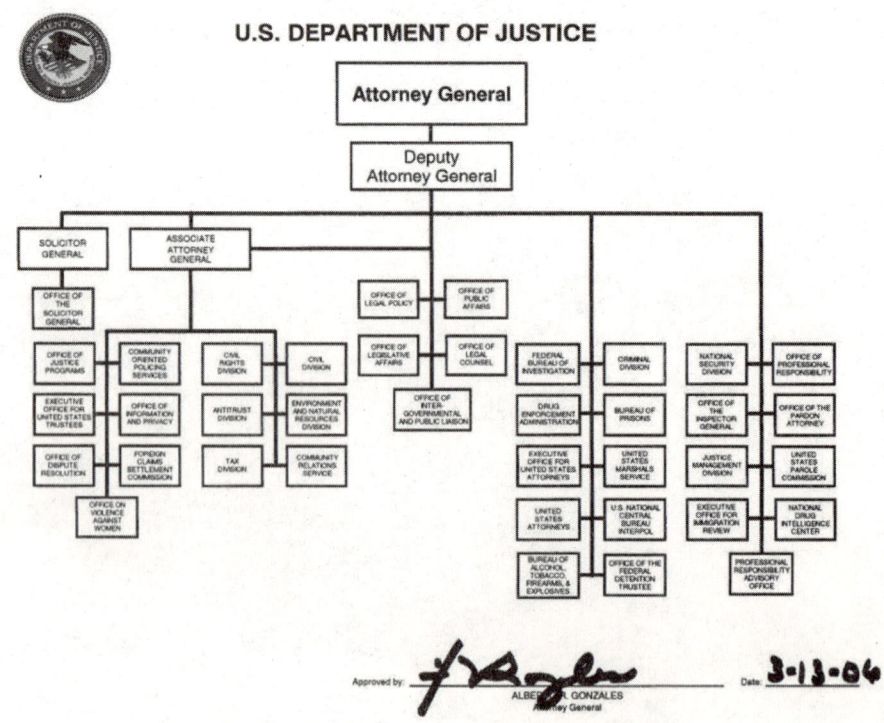

- **Flow charts**—show steps in a process.

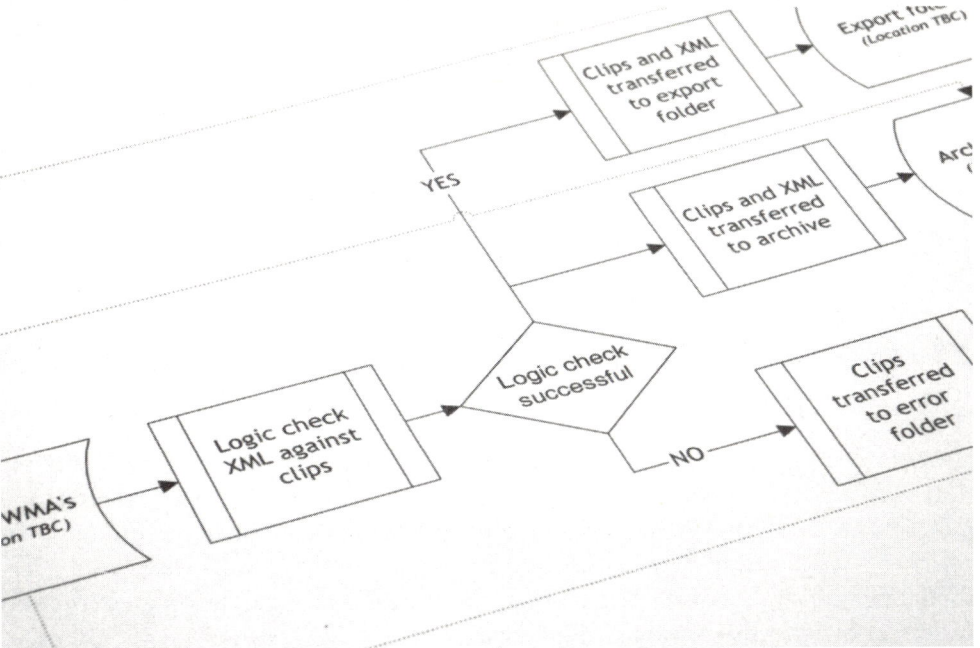

- **Pictogram**—a cross between a bar graph and a diagram, show comparisons using symbols.

Chapter 5: *Reports* 61

Keep in mind the different emphases of each chart type when deciding which to use. Which chart would be good for showing how users rated the difficulty of a task on a scale of 1 to 5? Which chart would help show the different levels of Internet use among users tested?

> For tips on creating charts see "Creating charts with Word and Excel"
> http://www.sandhills.cc.nc.us/english/wordguide/charts.html

Photographs, Illustrations, Diagrams, and Screenshots

Sometimes an exact representation is called for. Instructions often contain illustrations and diagrams (illustrations with labels) to help readers follow the written instructions and visually understand special relationships among parts (e.g., an arrow showing how to insert part A into part B). Diagrams can also be used to clarify the relationship of abstract concepts (example diagrams showing concepts and processes http://office.microsoft.com/assistance).

A screenshot, or screen capture, is an image of whatever is displayed on a computer monitor at the time of the screen capture (see Figure 5.3).

Figure 5.3. *Sample Screenshot. To make a screenshot, select the "Print Scrn" key on the keyboard, then paste the image into a Word file. Callouts (the red highlighting and explanatory text) help make a point*

Screenshots are helpful in usability reports when discussing faults with or making recommendations to improve an existing Web site interface. To create a screenshot of your Web site, open the desired Web page in a browser and click the PRINT SCREEN key. Then go to a Word file and click PASTE. You can resize and crop the image using Word's drawing tools. Screen captures can be turned into useful diagrams with the addition of *callouts*, which are arrows and explanatory text added to a screenshot. This can be done with screen capture software or with Microsoft Word's drawing tools.

> For more help with screenshots, see:
> - "Capturing screenshots" by ActiveWin.com
>   http://www.activewin.com/tips/mmedia_tips_9.shtml
> - "Shooting the Screen" by Rick Tew
>   http://www.efuse.com/Design/screenshots.html

Integrating Graphics within the Text

Visuals don't speak for themselves. This means that you must not leave interpretation of the visual up to the reader. You must reference the visual in the text and explain its *purpose* (what is it?) and *significance* (what is its main point?) to the reader. Create your graphics before you write your text and then, keeping in mind what is depicted in the graphic, write the explanation of the graphic. Use the same concepts and language depicted in the graphic for your explanation. *The visual always appears immediately following the paragraph where it is first referenced in the text.* For instance, you might write: "As Figure 5.1 shows, profit has decreased steadily since the merger," or, "Profit has decreased steadily since last quarter (see Figure 5.1)." Then you would insert the visual in the space immediately following this paragraph.

Since the best use of graphics is to clarify key points, always locate your visual as close to the relevant discussion as possible. While unimportant graphics can be located in appendices, this is not the best use of the persuasive power of visuals.

The design of graphics should be kept simple, uncluttered, and easy to read. Avoid overly complex three-dimensional visuals for this reason. If a reader cannot determine the point of the graphic in five to ten seconds, the graphic is not clear enough. Make sure to consistently re-size all visuals to fit within the margins of the page (e.g., all visuals 4 x 4" and center aligned).

Each chart or illustration is identified as a "Figure" and labeled consecutively. Each table is identified as a "Table" and labeled consecutively but separate from figures.

All figures and tables should be labeled with a caption that includes a number and a descriptive title that conveys the main point of the visual. You can also add a one or two sentence interpretation to the caption that further explains the graphic for readers. Place captions for figures below the graphic, and place captions for tables above the graphic (see Figure 5.4).

Figure 5.4. *Parts of the Caption. Locate the caption above tables and below figures*

Annotations on the figure:

- All graphics include a **caption** that includes the label, title, and description. Notice that for tables the caption is placed on above the table. For figures, the caption is placed below the —
- **Label** ("Table 1")
- **Title** ("Best Value Desktop PC")
- **Description** ("Dell offers…")
- If borrowing information or graphic, indicate **source** in fine print.

Table shown:

Table 1. Best Value Desktop PCs. *Dell offers the best all-around value. For a few more dollars, Gateway offers more speed and storage. Sony and iBuyPower: too pricey.*

Computer	Processor	Hard Drive	RAM	DVD	Price	Comment
Dell Dimension 4600	2.8-GHz Pentium 4 CPU	120GB hard drive	128MB NVidia GeForce FX 5200 graphics	16X DVD-ROM and 48X CD-RW drives	$1129	This well-appointed Dell offers fine performance for its configuration; an LCD and good speakers sweeten the deal.
Gateway 510XL	3-GHz Pentium 4 CPU	160GB hard drive	128MB NVidia GeForce FX 5200 G graphics	8X DVD±RW and 16X DVD-ROM drives, media-card reader	$1400	Nicely configured machine can handle basic text and spreadsheet documents, as well as multimedia presentations
Sony VAIO PCV-RS520	3-GHz Pentium 4 CPU	160GB hard drive	128MB ATI Radeon 9200 graphics	8X DVD±RW and 16X DVD-ROM drives, media-card reader	$1599	System with stylish design and competitive performance are undermined by inferior keyboard and tinny sound system
iBuyPower Back To School	2.2-GHz Athlon 64 3400+ CPU	80GB hard drive	256MB ATI Radeon 9800 XT graphics	8X DVD±RW drive, media-card reader	$1694	This system's high-flying performance makes it an appealing choice; outlandish front panel might give even gamers pause, though.

Source: PCWorld.com. Top 15 Desktop PCs. August 2004.

If you borrow information to make a visual, or borrow someone else's visual, identify the source in small print below the visual.

Adopt a Style Guide

Lastly, if you were writing a report for a large company, you would likely adhere to a particular style guide. Style guides dictate rules for proper punctuation, spelling, number usage, and page formatting. Since standards vary from company to company and field to field, it is important that writing teams decide ahead of time issues like how to spell non-standard words. For example, how will your team spell W-E-B-S-I-T-E? According to Associated Press style, it should be spelled "Web site." However, many in information technology fields spell it "website." The AP spelling is more recognized, but either version is acceptable, as long as the word is spelled consistently within the document. Style guides help resolve questions of style and mechanics during all phases of report production (drafting, revising, and editing).

ONLINE

Here are some sample style guides your team can adopt:

- U.S. Government Printing Office Style Manual http://www.gpoaccess.gov/stylemanual/
- Veridus Online Style Guide (Rocky Mountain Society of Technical Communication)
  http://www.stcrmc.org/Samples_and_Templates/LPStyleGuide.htm
- Rand Style Manual
  http://www.rand.org/services/pubsext/style_manual_ext/
- Microsoft Manual of Style http://www.winwriters.com/msmanualofstyle.htm
- University of Colorado at Boulder Style Guide
  http://www.colorado.edu/Publications/styleguide
- Resources for Documenting Sources by Purdue OWL (Good list of style guides arranged by field) http://owl.english.purdue.edu/handouts/research/r_docsources.html

Chapter 6

Resumes

According to one research report (Pibal, 1998), HR managers in the United States spend the following average amounts of time reading a resume:

- 1 minute   24%
- 2 minutes   31%
- 3–5 minutes   33%
- Other   12%

For these HR managers the length of time spent reading a resume varied according to the type of position, apparent applicable work experience, and resume quality. Many stated that poorly written resumes were immediately rejected.

What does this study suggest about writing resumes? For starters, your resume shouldn't be poorly written. The resume and cover letter are the first impression employers get of you. Careless typos or hard to read resumes won't do. Secondly, half of those reviewing your resume are going to spend less than two minutes deciding whether you make the cut, and most will spend under five minutes. Employers are reading closely and carefully—albeit quickly—to figure out if you fit their criteria for a particular job. If they don't see enough of what they're looking for, they'll simply move on.

Your immediate goal, then, is to convince employers to spend more time on your resume. (Your ultimate goal is to convince them to call you for an interview.) To succeed at this you need to be familiar with what most experienced resume reader/reviewers have come to expect from resumes. These are the rules, or *conventions*, of resume writing. You also need to be able to strategically utilize *page design principles* to create an initial and lasting impression with your resume. Most importantly, you need to be keenly aware of what that particular employer is looking for in an ideal candidate. Generic resumes cannot compete against resumes written by applicants who took the time to carefully tailor their job documents to the specific interests of the given audience.

Purposes

The resume introduces you to an employer in a short, quick way. It doesn't get you the job, but it can convince the employer that you're among the most qualified applicants and should be contacted for an interview. The resume then serves as a point of reference during the actual interview. Your resume, therefore, should:

- Be easy to read
- Establish your ethos as a qualified professional
- Provoke a desire for more information

Together with the cover letter, the resume is typically the first example the employer sees of your ability to communicate in writing—clearly, concisely, and neatly. The resume is an outline of your relevant skills. You cannot put your life's accomplishments into one or two pages, not now and especially not after you've been a working professional for several years. Your aim is to provide only the details that will convince an employer that you are qualified for the position. This usually means including details of your education, work experience, and anything else that shows you're qualified and capable. But you need to carefully and strategically think about what details should be included and what should be emphasized—and what should be excluded. You might start from a generic or general resume, but you should always carefully tailor it based on the reader's expectations according to your understanding of the job requirements and that company.

Format

Generally speaking, readers expect the resume to be one or two pages long. Entry-level resumes—from freshly minted college grads—typically do not exceed one page. But if you have extensive relevant accomplishments and skills from your college activities, you can consider two pages. The length of experienced professional resumes (with 5 to 10 or more years experience) varies according to the field, but most businesses still expect to see one- or two-page resumes.

Some basic format considerations include:

- **Paper quality**. Use good quality (20 lb. weight or more) white or off-white 8.5" x 11" paper. Laser quality copies only.
- **Font**. Limit the font styles to no more than two. Keep heading and body fonts to between 10 and 12 points per inch. Since you want your name to stand out you can get away with 14, 16, or maybe even 18 point font there. Keep in mind the look, feel, and impression that font styles make. Be conservative when selecting fonts. Your resume is not the time to dazzle with Matisse ITC—unless you judge it to be appropriate given your audience.
- **Balance**. Be sure your resume, as a visual unit, looks balanced across the four quadrants formed by the vertical and horizontal axes. Is your resume top-heavy, with too much text toward the top and very little toward the bottom? Is it left-heavy, with too much text toward the left margin? See the Visual Design section for more on this.
- **White Space**. This is a design term that refers to spaces on a page without text. You should always have a visually pleasing 1" border of white space around the resume. While you can fudge on this a little (e.g., 8/10" bottom margin), do not try to cram everything onto a page with 2/10" margins.

- **Chunking**. Break up or "chunk" your resume by adding white space after headings, between sections, and between items in sections. Is your resume so densely packed with text that it is difficult to read? If so, you should break up dense chunks with white space. Space consistently between and within sections. The spaces separating sections are usually larger than the spaces separating items within sections (HINT: You can select and then adjust the size of blank lines/spaces just as you can select and increase or decrease the size of fonts.)
- **Conventional Correctness**. Both the cover letter and resume should be free of mechanical mistakes and presentational errors: no misspellings, typos, strikeovers, or smudges. You should, however, understand that as an outline your resume does not have to adhere to the same rules for grammar and punctuation as prose documents (see the Style section below).
- **Headings and Lists**. Because your resume is an outline you should use headings and bulleted lists to help guide the reader.

Organization

The basic parts or sections of a resume include contact information, objective statement, education, relevant work experience or skills, other relevant information (like computer skills, honors and awards, or activities), and references. The basic rule of thumb when organizing information in your resume is to put more important information toward the top and toward the left of the resume. Western readers are conditioned to read left to right, top to bottom. Don't bury important details about yourself at the bottom of the resume. At the same time, resume readers have come to expect to see certain sections, like those listed above, so you must carefully work within those expectations but still try to take advantage of strategic organization to emphasize key information.

Novice resume writers are often unaware that there are some basic variations for organizing and presenting your information. Employment experts sometimes refer to these as resume *formats* or *styles*. These formats include the chronological, the functional, and the combination chronological/functional.

Chronological

This format organizes your work experience and education information in a straightforward reverse-chronological order (i.e., list more recent experiences first). For entry-level candidates this usually means education comes before work experience because their education is the most recent, most important relevant experience. For professionals with more time on the job whose work experience becomes more immediately relevant, education is generally moved to the bottom. This format is good if you have a strong history of relevant school and work experience. You basically organize your resume by where you worked/went to school or your job titles. The writer develops his or her relevant qualifications within the listing of jobs held. This style is less effective if you have unrelated work experience.

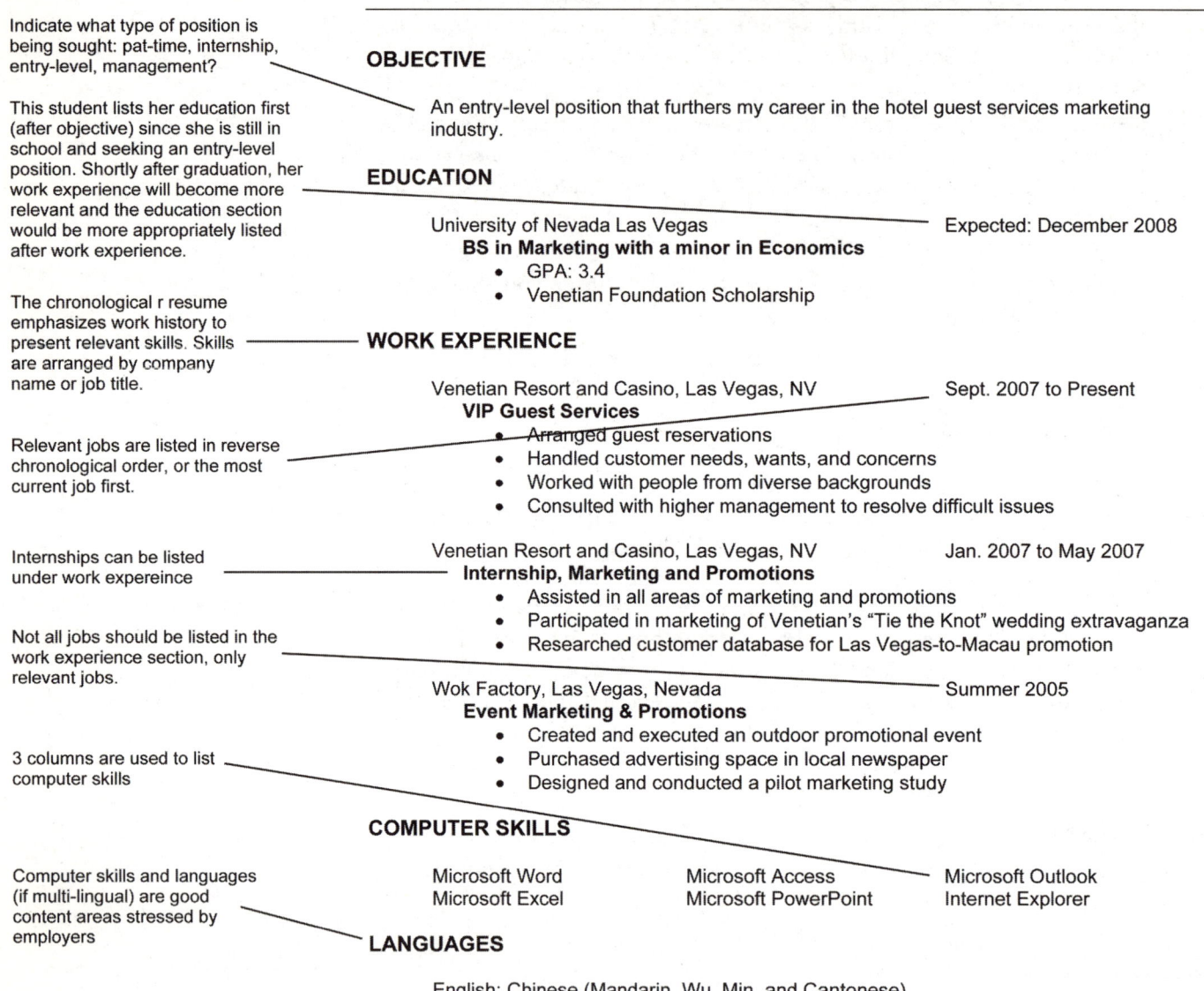

Figure 6.1. *Sample Chronolocial Resume*

Functional

This format emphasizes relevant skills, skills that are not necessarily related to jobs that you've held. Organize the resume by skill categories. Instead of a Work Experience section, functional resumes include Professional Summary, Professional Experience, or Relevant Skills sections. This format is a good choice for those whose work experience doesn't appear closely related to the targeted job, for those changing careers, or for those seeking a promotion. The functional resume elaborates on the required skills for a job. For example, if an internship asks for organizational, leadership, and communication skills you might consider developing evidence of your skills in each of these three areas.

A distinctive font can be used for the name. However, cursive fonts are harder to read and should be avoided for section headings.

A functional resume can still have an education section. This student is applying for an internship so her academic background is particularly relevant and thus is placed toward the top of the resume.

Listing courses is a good way to "fill" content on the resume, especially for sophomores and juniors who do not have a lot of work-related activities or experience. But employers assume that if you are in school, you are taking classes, so it is not necessary to list courses.

The functional resume emphasizes work-related skills arranged by type of skill or competency (not by place of employment).

Headings should highlight key skills highlighted in the job advertisement.

Relevant work history can still be listed in a functional resume but notice that no skills are development under the job, only place of employment is listed. The skills are developed in the functional skills section.

Combined headings allow you to merge related content areas that do not have enough content to be listed separately.

Cheryl Joyce Peters

2020 Crystal Lake Drive
Las Vegas, NV 89127

(702) 831-9035
peterscj@unlv.nevada.edu

Objective

An audit internship at Deloitte that uses my computer, organization, and communication skills

Education

University of Nevada Las Vegas Degree: May 2008
- Major: Accounting
- Minor: Finance
- GPA: 3.1

Relevant Courses

Financial Accounting	Cost Management
Managerial Accounting	Managerial Finance
Accounting Environment	Business Writing

Accounting Skills

Computer
- Use Microsoft Word and Excel at work and school daily
- Create income statements, balance sheets, budgets, cost analysis, and activity based costing using Microsoft Excel
- Make flyers and advertisements using MS Publisher

Organizational
- Compile data into reports and spreadsheets
- File client records and important documents
- Order supplies

Communication
- Writing letters, e-mails, and reports
- Communicate with employees and clients via e-mail
- Answer numerous phone lines

Work History

Peters Auto Body, North Las Vegas, NV 2002-Present
- Part-time bookkeeper, receptionist for family business

Memberships and Activities

- Member Beta Alpha Psi Fraternity 2007-Present
- American Society of Women Accountants, student member 2007-Present
- UNLV Student Government elected Senator for College of Business 2007-Present

Figure 6.2. *Sample Functional Resume*

Combination

The chronological/functional resume uses elements of both styles. In this format, the writer has to decide where he or she wants to develop relevant skills. You might provide an brief skills summary and then have skills developed in more detail under a Work Experience section, or vise versa; the writer may decide to provide a more detailed list of relevant skills, as in the Functional format, but include an abbreviated list of past jobs to show some employment stability, range of experience, etc. Choose the format that you judge best emphasizes your relevant skills and experience, but remember: *you can only develop relevant skills in detailed bulleted lists in either a work experience section or a functional skills section, but not both.*

Also, keep in mind that while there are common conventional headings like Education and Work Experience, you should choose section headings that help emphasize your relevant skills (like the functional internship resume example above). Likewise, if you don't have any noteworthy honors and awards, then don't include such a heading.

Betty P. Thorne

3115 Oliver Street
Las Vegas, NV 89134
(702) 234-5678
thorneb@yahoo.com

OBJECTIVE

Seeking an exciting internship in a vibrant and fast-paced advertising environment.

SUMMARY OF QUALIFICATIONS

- Majoring in Journalism and Media Studies at UNLV
- Over six years experience in customer service
- Strong written and verbal communication skills
- Attended National Writers Workshop in 2007

EDUCATION

University of Nevada Las Vegas, Las Vegas, NV   Degree: May 2009
Major: Journalism and Media Studies
- G.P.A: 3.0

University of Brighton, Brighton, England   Jan. '06—June '06
- Semester Abroad G.P.A: 4.0

EMPLOYMENT HISTORY

Canyon Ranch Spa Club at the Venetian, Las Vegas, NV   Oct. '05—Present
Reservations Specialist/Guest Relations
- Greeted guests, scheduled appointments, and responded to customer needs
- Replied to guest queries via e-mail and letter
- Composed thank you and letters and other guest correspondence

Color Me Mine: Paint Your Own Ceramics, Las Vegas, NV   Oct. '02—Jan. 05
Assistant Manager/Customer Service
- Interacted with customers and supervised part-time staff of 2-3 employees
- Maintained daily sales receipts and customer transactions
- Used Excel to track inventory and sales records

RELATED ACTIVITIES

National Writers Workshop, Fullerton, CA   May 2007
- Attended seminars in copywriting and advertising writing

Side annotations:

This combination resume example includes an brief summary of relevant qualifications, which is a snapshot or highlight of the candidate's main qualifications and literally a summary of content listed in the rest of resume. Make sure the summary list corresponds to key details developed elsewhere in the resume.

This writer has chosen to develop skills under the work history section of the resume. One could ask if these skills are relevant enough for an advertising internship to warrant development in a work history section. Perhaps a functional resume would have been a better choice for this student.

Here are some key relevant experiences buried at the bottom of the resume. How could this writer rearrange the resume to highlight this experience?

Figure 6.3. *Sample Combination Resume*

Content

Content refers to what details you choose to include and how you choose to represent them. The resume usually provides the following information:

Contact Information

Include your name, address, telephone number, and e-mail address. Sometimes students include both their current or college address and their permanent or home address. You will need to decide if this extra detail is necessary. If you're applying for a job at the beginning of the semester, is it likely employers will need to contact you at your parents' house? List only the information that will make it easiest for the employer to contact you, so use one phone and one e-mail address preferably.

You can add a graphic element such as a horizontal line to visually separate your contact information from the body of your resume. Some resumes include a simple image or design for visual appeal.

You can also use the same design and layout of your contract information on your resume and cover letter to give both documents a more professional and consistent visual design.

For resumes indented for print reading only, turn off MS Word's automatic hyperlink function when writing your email address. Highlight the address, right-mouse click, and go to hyperlink in the menu, then select "turn off."

Objective Statement

The objective statement includes 1-3 lines of text, summarizing the position(s) you are applying for and/or your main qualifications. In keeping with the principal that resumes are outlines, your objective statement should be written as a complete sentence but rather a phrase (see examples below).

Employers need to know what job you want (e.g., "an entry-level finance position" or "a Support Specialist position") and either what your goals are (e.g., "in a top-five accounting firm") or what your qualifications are ("that utilizes my computer science and management information systems skills"). Don't be too vague (e.g., "An internship allowing me to utilize my knowledge and expertise in different areas")

There are four main approaches when writing a good objective statement:

1. Emphasize a specific **position** and your main relevant **qualifications**	"A position as a Support Specialist allowing me to use my skills in the fields of computer science and management information systems."
2. Emphasize the **field** or type of **organization** you want to work in and your main **qualifications**	"To join an aircraft research team allowing me to apply my knowledge of avionics and aircraft electrical systems."
3. Emphasize your **professional or career goal**	"To help children and families in troubled situations by utilizing my child protection services background."
4. Emphasize a **specific position** desired	"Technical writer specializing in user documentation."

Source: "Writing the Objective Statement for Your Resume." Purdue OWL.
URL: http://owl.english.purdue.edu/workshops/hypertext/ResumeW/objective.html

Education

Include name of college, degree, major, and date of graduation, formatted in the same order as your work experience (e.g., school is equivalent to company name, major is equivalent to job title, etc.) You don't have to include your GPA, but employers will assume it is low if it is missing, so include even "average" GPAs. You may also list significant courses or academic honors in this section. Arrange multiple schools in reverse-chronological order. For entry-level candidates, education tends to be one of the most important qualifications, so it is listed toward the top of the resume. Working professionals with several years of experience tend to move their education to the bottom of the resume.

Experience

Include your position/job title, name of company, address, and dates of employment. Include, but do not emphasize, irrelevant work experience. Provide detailed bulleted action lists that evidence your relevant qualifications, skills, and accomplishments. Be very specific and concrete. You should consider which skills from that job are most suited to the job you're applying for. Instead of a generic list of eight skills, choose the four that best match the employer's criteria. Don't just cut-n-paste your job duties, instead list particular accomplishments/achievements. Use industry jargon and numbers/statistics to list specific projects you worked on and productive results. Do not overdevelop the obvious. If your job is Pizza Delivery Person, you don't need to list delivered pizzas as a skill. Make sure your lists are parallel in structure. If you start your job descriptions with a past-tense verb, then you should start all job descriptions with a similar past tense verb (e.g., *created, sold, supervised*). Remember, you can develop your experience chronologically or functionally, or combined.

Special Skills, Honors, and Activities

Along with your education and experience, you should consider what other curricular or extra-curricular skills, activities, or accomplishments help establish your qualifications and ethos. It is often a good idea to include a separate computer skills section. Employers want to know how you can contribute to the particular position you're applying to. They're very interested in your career-related skills, academic- and community-oriented activities, and special aptitudes. Being involved in a range of activities helps show that you can balance a busy schedule and are not one-dimensional. Employers are less interested in your hobbies or outside interests, however.

If you're involved in college athletics or social organizations (e.g., Greek fraternities or sororities), make sure you don't overemphasize this. Executive positions held in social organizations, like being President or Treasurer, might help show relevant leadership or financial experience, but don't list every position you've held in your sorority for the last four years. Employers generally want to know what you've done lately, so if you're a junior or senior in college, this usually means that your high school accomplishments aren't that valuable.

References

It's usually taken for granted that you have a list of references prepared if employers request it. You shouldn't list references and contact information in your resume. You don't even need to have this section, unless you maybe need to fill some empty white space on your page (e.g., "References – Available upon request").

Style

Style refers to the accepted ways to present resume content. Remember, above all, that the resume is an *outline*. As such, you do not need to conform to the same grammatical and mechanical principles as prose documents. The following guidelines help you present your information clearly and quickly:

- Do not use complete sentences and avoid the first-person "I." Short, concise fragments with action words are easier to read than lengthy prose.
- Limit the use of punctuation. You should still use commas to separate items in lists, but do not use end-line punctuation. You can eliminate punctuation in most common abbreviations, e.g., GPA instead of G.P.A.
- Use lists to make information easy to scan. Break up lengthy sections of dense, packed prose into bulleted lists of parallel grammatical structure.
- Arrange more important information first. Prioritize items in lists for example.
- Use technical or industry jargon to help establish your professional ethos. Employers look for keywords to help them decide what type of position you can fill in their organization. Use as many nouns as you can that pertain to your desired field of work.

- Quantify your experience wherever possible (e.g., "supervised 15-member staff," "increased profits from first to fourth quarter by 70%")
- Stress your strengths. Don't mention weaknesses.
- Be completely honest. Do not misrepresent or fabricate skills. Your resume will be used to generate interview questions, so don't put yourself in a position where you'll have to deny anything stated in the resume. You won't get the job. You can even be banned from UNLV campus recruiting if caught.

Resume Visual Design

Many of the techniques for making your document easier to scan fall into the realm of page design. Because the average HR manager spends one to three minutes making initial decisions about the resume you need to ensure that your resume will not frustrate the reader and, indeed, that it will invite them to actually spend more time on it. The following page design principles used by technical writers, designers, and desktop publishers will help you do this:

Create Visual Balance

A page can be divided into four quadrants, as if folding the resume in half twice. Make sure that text appears in each section relatively equally, which is known as symmetry. If you want to create a purposeful asymmetry, fill the right top and bottom quadrants with more text.

Image source: http://owl.english.purdue.edu/workshops/hypertext/ResumeW/wholepage.html

The "quadrant test" is a way to evaluate the visual balance of your resume. Fold your resume draft into quarters (as indicated in the picture above). If there is too much empty white space in the bottom two quadrants, then you need to add more content to your resume. If there is too much empty white space in the top right and bottom right quadrants, then it usually means you need to expand the detail in your bulleted lists (do not have short, 3-4 word bulleted lists of skills). Also notice if you have a consistent 1" border of white space around all four quadrants. You must balance your resume if ANY border is larger or smaller than 1".

Create Visual Hierarchies

Designers use headings, indentation, and highlighting (i.e., changing the size or format of text) to create noticeable levels of information. You can signal major section headings by bolding text, for example. You can then signal different parts within sections by indenting and italicizing the next level of information. This creates contrast by creating hanging *indentation*.

Level of Information	Formatting	Result
Level 1: Section	13 pt. Arial, bold	**Work Experience**
(White Space)		
Level 2: Company	12 pt. Times, bold, indent	**Acme Finance**, San Diego, CA
Level 3: Job title	12 pt. Times, italics, indent	*Market Analyst*
Level 4: Relevant skills	12 pt. Times, indent, bullet	• Analyzed… • Prepared… • Managed…

Visual hierarchies, i.e., clear levels of information, are created by font style, highlighting, and indentation

You can also use tabs or borderless tables to put lists of information into more readable columns.

Instead of:

Related avionics courses
Aircraft Electrical Systems, Aircraft Avionics Systems, Digital Electronics, and Optical Physics.

Other related courses
Computer Programming, Multivariate Calculus, Advanced Composites, Fiber Optics

Chapter 6: *Resumes*

Use cleaner columns:

Coursework

Avionics Courses	**Other Courses**
• Aircraft Electrical Systems	• Computer Programming
• Aircraft Avionics Systems	• Multivariate Calculus
• Digital Electronics	• Advanced Composites
• Optical Physics	• Advanced Composites

Use Highlighting

Highlighting refers to any manipulation of plain text for emphasis, including bold, italics, underlining, all capital letters, or altered size. Choosing a different font style or using a larger font size to create contrast for headings is an example of highlighting. Avoid overemphasizing text; you don't need to simultaneously bold, italicize, and underline your major section headings. Use highlighting mostly to distinguish levels of text (i.e., for headings and subheadings). Do not use more than two font styles in your resume.

Format Consistently

Be sure to follow the principle of consistency when creating visual hierarchies and levels of emphasis. Use the same amounts of indentation and spacing for the same levels of information. Use the same font style for the same levels of information and so on. One way to test for consistent design is to use the vertical columns test. Use imaginary vertical lines to check that information of the same level aligns visually in the same vertical column.

Evoke Simplicity

Create a clean, professional look to your resume. Do not use more than two font families; choose conservative fonts; and do not overemphasize text. Be sure to balance text and white space on the page. Do not pack the page with too much text, but don't have too much white space either.

Remember that design principles are rules-of-thumb, not absolute laws. You'll have to use your judgment about what looks best based on the above principles, careful study of models, feedback from experts (Career Services Center staff, business professionals, etc.), and feedback from family and friends.

Scannable Resumes

Many mid- to large-sized companies use electronic resume management systems (ERMs), sometimes called applicant tracking systems. These software database applications allow the employer to **scan** a print resume into a computer database that the employer can customize to search for certain key terms. The ERM system thus sorts and ranks resumes in its database based on the employer's predetermined criteria. So, in some companies, your application is initially being screened by a computer! How many companies use ERMs? It's not known for sure, but according to Proven Resumes.com, since ERMs cost upwards of $100,000, only companies that hire lots of employees will use applicant tracking systems. Most technology companies and many Fortune 500 companies use ERMs. If your job advertisement specifically asks for a scannable resume, be sure to format accordingly. If you're unsure, you can either make your print resume scanner friendly, just in case, or you can correspond with the company beforehand.

The latest ERMs can scan most any print resume without confusion or data loss. However, there are some things you can do to ensure that your resume is scanner friendly:

- Mare sure no characters/letters touch each other
- Use sans serif fonts (Arial, Helvetica, Verdana, Optima) rather than serif fonts (Times New Roman, Courier)
- You can use italics and bold, underlining, and horizontal and vertical lines as long as the letters don't run into each other
- Do not use ampersands, percent signs, or foreign characters because they may not translate
- Use paragraphs, not columns, to list information, as the scanner reads text from left to right
- Use white paper and do not fold or staple your resume if mailed; ERMs don't count page numbers so you can get away with a longer scanner friendly version of your resume
- Make sure you have keywords located throughout your resume to increase the number of "hits" during the employer's database search. You can also add a **keyword summary** that frontloads and repeats keywords located throughout the rest of your resume:

KEYWORD SUMMARY

Advertising, Copywriting, Marketing, Public Relations, Creative, Design, Marketing, Journalism, Media, Fashion, Comedy, Campaigns.

Conclusion: Do it Your Way!

Remember that conventions for resumes vary from field to field. One strategy for learning and adapting to context-specific conventions is by studying models. Bookstores stock dozens of resume how to books with hundreds of samples.

> Google has a directory of resumes listed by industry that you can use to see how people in your field present their skills. URL: http://directory.google.com/Top/Business/Employment/Resumes_and_Portfolios/By_Industry/

It is perfectly okay to collect and study samples of resumes and cover letters. (Just be careful when looking at web resumes. Most people essentially post their print versions on the web with little modification, but there can be differences between print and web documents, like hyperlinks). But you must imitate models with a critical eye, keeping in mind *your* skills and *your* particular audience. You should never make random or uncritical choices, like plugging your skills into someone else's template. What if the sample you're imitating is in chronological format when the functional format could better represent your skills? How can you accomplish your goal of creating a distinctive looking resume that sets you apart from others if you rely on a Microsoft Word resume template? Like writing any document, you need be able to justify your content and design choices based on document conventions and document design principles.

When it comes to writing resumes, "doing it your way" really means making strategic choices within the boundaries of convention. You need to know enough about the rules to be able to bend them without breaking them.

End Notes

[1] Garcia, Gary. (1994). "Job Search Documents: The Resume." *Business Writing Coursepack*. Purdue Business Writing Program. Copymat: West Lafayette, IN.

[2] Kopp, Brian M. (2001). "Online Workshops: The Resume" Online Writing Lab. Purdue University. http://owl.english.purdue.edu/workshops/hypertext/ResumeW/index.html

[3] Pibal, Darlene. (1998). "The Importance of Resume and Cover Letter Criteria in the United States, Australia, Germany, Thailand, and Hong Kong." *Delta Phi Epsilon Journal*, 40.3, 137-143.

[4] Pollock, Ellen Joan. (2005). "Is Your Resume Ready to Be Scanned?" *The Wall Street Journal Online.* http://www.careerjournal.com/jobhunting/resumes/19980730-pollock.html

[5] "Scannable Resumes." (2004). ProvenResumes.Com. http://www.provenresumes.com/reswkshps/electronic/scnres.html

Chapter 7

Cover Letters

Your cover letter and resume should work hand-in-hand to present your qualifications to an employer as completely as possible. Your resume is a list or outline of your most relevant qualifications, while the cover letter highlights the specific skills or attributes that qualify you for the position. Your cover letter should not be a restatement of your resume. Instead, it should clarify for the reader how you meet the qualifications identified in the job advertisement.

Aims of the Cover Letter

Employers want to get as much information as possible before making decisions about whom to interview. They don't want to interview unqualified candidates or candidates with padded resumes. The cover letter gives them one more opportunity to evaluate the quality of their candidates. However, this should be good news to you, for the cover letter gives you one more opportunity to connect your attributes with the needs of that particular employer, one more chance to be persuasive. Of course, to write a truly effective cover letter you need to research the company in some detail and show readers that you have a working knowledge of their organization.

Create a Professional Ethos

Researching a job demands a commitment of your time, but it allows you to craft a cover letter that provides a unique introduction of yourself. The cover letter may be your first contact with this employer so you want to be sure that you make a favorable first impression. Your cover letter needs to look like a formal letter and follow appropriate conventions. Pay attention to form and spacing so that your letter is aesthetically appealing.

Moreover, you need to avoid spelling, punctuation, and grammatical mistakes. A poorly written letter can reflect badly on you as a person and as a potential worker, since it implies that you are not interested in taking the time necessary to produce a document of quality. Avoiding errors is hard work, so take your time and check the entire letter closely. Read and reread your letter, and get others to read it. Also, don't be dependent on spell-checkers and grammar-checkers. Spell-checkers, especially, can allow typos to inadvertently slip through.

Your cover letter must create the impression that you are knowledgeable about the company, the job, and your profession. You should be able to write authoritatively about all aspects of the company, especially those that are most important to you. You should be able to discuss in detail how your past experiences enable you to perform the duties described in the job advertisement. To establish your credibility, use keywords mentioned in the job description and drawn from your major or professional field.

Persuade Through Examples

The main goal of the cover letter is to convince the employer that you can do the job, that you have the qualifications desired by the employer. The cover letter should not repeat the resume, but instead emphasize those elements of the resume that are the most important. Read the job advertisement closely to identify the most important qualifications. Arrange the cover letter to emphasize your two or three most relevant qualifications. For example, if the advertisement calls for knowledge of human resource management, three years experience, and strong communication skills, then address how you meet each criteria in three separate paragraphs in the body of your cover letter.

But you can't just claim to possess a certain qualification; you must **demonstrate** your claims with evidence, specific examples that convince the reader you indeed have the desired skill. Instead of writing, "I have tournament planning experience," write:

> *I have tournament planning experience. In the summer of '96 and '97, I was a little league coach. For those two summers I prepared the round-robin brackets for a 10-team regional tournament. I learned how to set up the brackets, make sure they were fair, and declare an undisputed winner.*

The cover letter allows you to provide specific examples that show the reader your skills and discuss relevant qualifications in more detail than what is merely listed in the resume.

Show Your Interest

A cover letter should never read like a generic form letter. You can write a general resume for use at job fairs, but a cover letter cannot be used in the same fashion. You need to show readers that you are aware of their needs and that you took the time to write a letter directly to them. While the cover letter and resume are about you, they also must be tailored to show the reader how important the company is to you as a prospective employee. But avoid peppering your cover letter with a lot of needless flattery.

If you have done the necessary research, you should be excited about the possibility of working for this company. Show them your knowledge of the company and that you want to be a part of the company. Point out specific features of the company that attract you. A prospective employer can't help but feel pride when reading about the positive attributes of her or his company. Your cover letter should convey your desire to work in your particular field of interest, so you need to express this enthusiasm as completely and sincerely as possible.

Express Your Personality

Another reason for writing a cover letter is to express your personality more fully than you can in a resume. Since the cover letter is your primary introduction, you need to work hard on your writing style. You should write your letter in your own words, avoiding inflated language or words that you are unfamiliar with. Try to be natural and use the language of your field to show your expertise. You don't need to sound more intelligent by using impressive sounding words. In fact, using words that you are unfamiliar with will often leave readers unimpressed, especially if you use words inappropriately. You should, however, be able to use jargon appropriate to your field of expertise, but don't go overboard.

While you want to express a confidence in your abilities, avoid self-absorption. Unqualified, grandiose statements using words like *ideal* or *most qualified candidate* tend to leave the reader unimpressed. Also, don't let humor invade the letter. You have no idea if the reader shares your sense of humor.

Cover Letter Format

Cover letters include the traditional features of a formal letter (heading, salutation, opening, body, closing, signature block, and end notation) and most are limited to **three-quarters to one full page**. You can adjust the point size or the font so that the letter is visually appealing (you can balance the letter on the page, as with resumes, usually by vertically aligning the page). You need to be both succinct and detailed. Provide specific examples using lots of concrete details. Most importantly, use your own words. Do not copy verbatim from samples or create a letter based on a formula.

Annotation	Letter
Include a subject line that references position title and number, if given.	
Use a formal salutation.	
In the opening, state what job your are applying for and how you learned of it. Explain your motivation for applying and summarize the key skills you will develop in the body of the letter.	
In the body, provide 2-3 paragraphs that demonstrate how you meet the qualifications.	
Limit each paragraph to a single skill or set of attributes, indicated by the topic sentence of each paragraph. Provide concrete details to convince reader that you possess the skills indicated.	
Relate past experiences to company needs.	
Internship applications can include a paragraph explaining how your see the internship furthering your career objectives.	
In the closing, politely request an interview or state a specific time that you will contact the employer. Provide contact information, offer to answer questions, and thank the reader. Maintain a courteous tone.	
Include end notation.	

3456 Crystal Waters Street
Henderson, NV 89052

March 27, 20XX

Allison Hess, Public Outreach Director
Ecolutions
10271 W. Pico Blvd.
Los Angeles, CA 90004

RE: Public Relations Internship

Dear Ms. Hess:

I am writing to apply for the Public Relations internship advertised on Wetfeet.com. I admire your company's interest in educating young children about environmental issues. I believe this is key to reaching your long term goal of pollution prevention. As you can see from my attached resume, my education, work experience, and interest in the environment are well suited to this position.

I am currently a junior at University of Nevada Las Vegas with a 3.4 GPA. I am pursuing a dual major in marketing and environmental studies and have taken several upper-division marketing classes. I am a member of the marketing club and recently was awarded a scholarship by one of Las Vegas' top advertising agencies based on a portfolio of my marketing course work.

My work experience has developed my communication and marketing skills. As a legal assistant for Jamison, LLP, I have drafted numerous letters and legal documents. I also answered phones and performed other office duties. As a caricature artist, I used direct sales and a marketing packet that I prepared to become a top seller among artists in my group. These skills will apllow me to assist in the preparation of Ecolution's media campains and event plans.

Perhaps the most important attribute that would help me succeed at Ecolutions is my passion for the environment. A few years ago, I learned about the devastating effects that livestock production had on the environment, and I became a vegan. It has always frustrated me that this is only a small solution. This is why I decided to pursue a major in environmental studies, so I could turn my desire to improve the environment into a career.

It would be a privilege to intern at Ecolutions. I would like to have the opportunity to speak with you in person about my application. If you have any questions or wish to schedule an interview, please contact me at (702) 612-7823 or at sue888@hotmail.com. Thank you for considering my application.

Sincerely,

Sue Smith

Encl: Resume

The Heading

Whenever possible, identify an actual contact (or hiring manager) within the company and address the letter to that person by name. Your chances will improve if you can find out through research exactly who makes the hiring decisions. Use a formal address (Mr., Ms., Mrs., or Dr.) and be sure to spell the name, title, company, and address correctly. Include a bolded subject line about the position before the salutation.

The Opening

The introduction of the cover letter

1. identifies a specific job and how you learned of the position,
2. persuasively explains your motivation for applying to the position
3. summarizes the skills you will develop in the body of the letter

The reader should know immediately why you are writing this letter. Don't waste time with clever anecdotes, and don't make the reader guess what it is that you want. You need to state the specific job that you want and state specifically how you became aware of the position. You then need a persuasive statement that explains your motivation for applying, followed by a brief list of your key skills or a preview of the attributes that the letter will discuss. In this way you should be able to clearly connect your skills with the needs of the company and establish how you will address the qualifications outlined in the job advertisement.

The Body

Body paragraphs should always begin with a **topic sentence**, or a brief statement of the skill emphasized in the paragraph. While cover letters are normally read much more slowly and much more closely than resumes, a topic sentence allows readers to scan the letter easily as well and still get the majority of the information that you are trying to convey. Using topic sentences also helps you stay focused on one idea per paragraph. Readers always have difficulties when paragraphs try to cover too much ground or deal with too many points. Before you begin drafting, create a topic sentence outline, arranging the topics according to their relevance (i.e., discuss your most relevant qualification in the paragraph immediately following the opening).

Body paragraphs should above all include specific details and examples that support any claims about your relevant skills made in the topic sentence. The examples in your body paragraphs should refer to information specifically requested in the job ad. This does not mean parroting to the readers the job ad, but *showing* them your qualifications through examples.

It is a good idea to refer to your resume at least once in the course of your cover letter. You want to convince the reader to read your resume and look at specific areas of your resume that they will find particularly interesting.

The Closing

Finally, your conclusion should tie up any loose ends and let the reader know what's next. You should either request an interview or state a specific time when you will contact the employer about your application. The conclusion should also provide contact information (phone and e-mail) so that the employer can contact you quickly and easily. You can refer to the address at the top of the page, but it is advisable to provide readers easy access to contact information in the closing, as well. Maintain a courteous tone, thank readers for their time, and offer to answer any questions.

Should You Submit an E-mail Cover Letter?

Sending your application by e-mail is faster, a good way to show off your technology skills, and preferred by most executive recruiters. If the job ad directs you to submit your application by e-mail, include your cover letter in the text of the e-mail message. Don't be afraid to send the exact same information via both snail mail and e-mail. However, experts agree that the e-mail cover letter should be shorter than its print counterpart, because readers are impatient with long, scrolling e-mail messages. You can compose your e-mail cover letter off-line using a word processing program, then use your software's copy and paste tools to copy the text into the e-mail message.

Include a subject line that references position title and number, if given.	**To:** A_Hess@Ecolutions.com **Cc:** **Subject:** Public Relations Intern **Attach:** Smith_PR_Intern_resume.doc (23.5 KB)
Attach a copy of your resume. Give the file a specific file name that the reader can recognize among other resume files.	
Include a formal salutation.	Dear Ms. Hess:
In the opening, reference your attachment, giving the format and name of the file. Offer to provide an alternative format if desired.	I am writing to apply for the Public Relations internship advertised on Wetfeet.com. I believe you will find my education, work experience, and interest in the environment well suited to this position. I have attached an MS Word XP version of my resume (file name: "Smith_PR_Intern_Resume"). If you have any problems with the attachment or require a different format, please contact me.
Shorten the contents to roughly **half** that of a print cover letter.	I am currently a junior at University of Nevada Las Vegas (3.4 GPA) pursuing a dual major in marketing and environmental studies. I recently was awarded a scholarship by one of Las Vegas' top advertising agencies based on a portfolio of my marketing course work.
If your e-mail application doesn't have spelling or grammar checkers, copy the text from a word processing application.	My work experience has developed my communication and marketing skills. I have drafted numerous letters and legal documents as a legal assistant and used my promotion skills to become a top selling caricature artist. These skills will apllow me to assist in the preparation of Ecolution's media campains and event plans.
Keep the closing similar to a print cover letter.	I am interested in a career that will allow me to protect the environment and would be honored to intern at Ecolutions. I would like to have the opportunity to speak with you in person about my application. If you have any questions or wish to schedule an interview, please contact me. Thank you for considering my application. Sincerely,
Include your address and contact information in the signature block.	Sue Smith 3456 Crystal Waters St. Henderson, NV 89052 (702) 612-7823 sue888@hotmail.com

Copyright © 2004 Kendall/Hunt Publishing. All rights reserved.

As for how you submit the resume electronically, most employers will accept resumes sent as Word file attachments with the file named after the candidate (e.g., Sue_Smith_resume.doc). Some employers, wary of viruses, will ask for a plain text version of the resume included in the body of the e-mail. This is often referred to as an ASCII resume, named after the plain-text file format. (See Kennedy[1] for how to format the plain-text resume.) If no instructions are given in the job advertisement you can include both an ASCII version in the e-mail and attach a Word file version. By doing both you won't have to worry about whether the employer wants to download your attachment.

In the opening of your e-mail cover letter you should mention the format in which you've submitted your resume. You could also offer to send another resume if the employer prefers a different format. Finally, since the e-mail cover letter doesn't contain a heading with your contact information, you should include your address, phone number, and e-mail address after your name in the signature block.

End Notes

[1] Kennedy, Joyce. (2003). "Fast Key to ASCII." CareerJournal.com. URL: http://www.careerjournal.com/jobhunting/resumes/20030909-kennedy.html

"Cover Letters" contributed by Ed Nagelhout

Chapter 8

Presentations

Most oral presentations in business settings, except for formal presentations such as a CEO's keynote address, are extemporaneous. That is, you need to plan and practice your presentation, but the actual delivery needs to appear as if you are talking off the cuff. *Talking* is the keyword. Business presentations, whether formal or informal, are not read verbatim from 3 x 5 cards. They are not even memorized and repeated word for word. In business settings, where you never want to lose your audience because of the monotone recital of some canned speech, presentations need to be lively and interactive.

In formal presentations you'll likely be selling some idea or product to a small group in a confined conference room, so establishing a positive, personal relationship is critical. In informal presentations where you know your audience more closely, you might be informing supervisors and/or coworkers of the status of a project. In either case, reading to your audience won't do. You need to establish a knowledgeable, confident, and prepared ethos during business presentations. Relying too much on notes suggests that you don't know what you're talking about or didn't care enough to prepare beforehand.

Instead of reading verbatim from notes, you need to talk to your audience, make eye-contact with them, and draw them into your presentation. To help guide you and your audience, you should use *visual aids*, including slides, print literature (handouts), computer demonstrations (projected onto a screen), whiteboards, flip-charts, etc. You should also make your audience active participants rather than passive recipients of your message by asking questions, elaborating with easy-to-understand visual examples, and perhaps adding hands-on activities.

Purpose

Be sure to understand up-front what the desired outcomes of your presentation should be, or what the audience should know about or know how to do at the end of your presentation. Decide what the main point of your presentation should be and define what action you want them to take afterward. Do you want your audience to understand why your product is best for their needs? Do you want them to abandon the planned merger? To buy your product? To know how to use the company's new voice-mail system? Build your whole presentation around getting your point across persuasively.

Audience

Be sure to analyze and define your audience—who they are, what their relationship to you is, what their primary occupations, activities, values, etc. are. Define the context of your presentation as well. Where is your audience working/learning? How will they benefit from your presentation? What do they need to know to use your information effectively?

Be sure to also define your audience's existing level of knowledge about your presentation topic. Consider how well they know the topic you're going to discuss. What can you assume they know? What concepts and terminology will you need to explain, define, spend more time on, etc.? A presentation on how to use Microsoft Outlook will have to be tailored depending on whether the audience uses Outlook regularly and wants to learn more about advanced features, has never used Outlook, or has never used a computer.

Organization

The structure of effective presentations dates back to Classical times. Aristotle wrote about it in the fifth century, B.C. To get your main point across you need to build into your presentation *planned redundancy*. That is, first tell your audience what you're going to tell them, then tell it to them, and when you've finished, tell them what you just told them. Sound familiar? Of course, it's the standard introduction-body-conclusion structure. It applies doubly to presentations, where your message will be competing with your audience's short, wandering attention-span.

The standard organizing pattern of presentations is as follows:

- **Introduction**: Introduce yourself to your audience and establish your presentation's purpose and/or outcomes. Mention your name, some background (e.g., organizational affiliation, if you have one, or your major), and your reason for talking. Establish your credibility with a crisp, practiced opening that gets your audience's attention and gives a clear sense of direction. This is called signposting, or giving your audience a roadmap for following your presentation.
- **Overview**: You may decide to start out with some attention-getting opening before or after your overview, but at some point after introducing yourself and your topic and before you start the main part of your presentation, you absolutely must "signpost," or outline, the main points of your presentation.
- **Discussion/Body**: Cover the main points of your presentation in a logical, easy to follow order. You can lend coherence to your discussion by including clear transitions (e.g., *next, secondly, another, Now Jane will...*, etc.).
- **Conclusion**: Tell them what you told them. Give your audience a clear signal that that the presentation is complete by including a concluding or closing discussion that reiterates the main points.
- **Q&A**: You should build time into your presentation for questions and answers. Anticipate and invite questions.

Research suggests that a typical person's attention span lasts anywhere from 10 to 20 minutes. You know this from sitting through boring lecture classes. To secure your audience's attention and help them remember whatever idea or product you're selling, be sure to include *interactive elements* in your presentation. These include such techniques for keeping your audience's attention as the following: questions for them to ponder and maybe even answer; examples, anecdotes, and stories for them to visualize; references to visual aids like charts and figures or handouts; actual activities or demonstrations for them to participate in; and even (appropriate) jokes.

Visual Aids

A visual aid is anything used to enhance the oral delivery of your presentation including a PowerPoint slideshow, overhead transparency, flip-chart poster, handout, or physical object. Commercial Web sites such as Presentations.com and PresentersOnline.com offer expert advice for choosing and effectively using the range of visual aid technologies available. Presentations are increasingly being delivered over the Internet and through real-time web, video, and telephone conferencing applications. Using presentation software applications like PowerPoint, Corel Presentations, or PowerJam Studio to aid the delivery of presentations has become standard, though you need to carefully decide what visual aids will be most appropriate, and feasible, for your particular purpose, audience, and context.

When using presentation software to create visual support for your presentation, as a rule of thumb keep the design of your slideshow *simple* and *consistent*, and *appropriate* and *appealing* to the audience. In "Power Pointless," Rebecca Ganzel discusses how many workers overzealously try to integrate all the bells and whistles of PowerPoint. Workers get wrapped up in working on the presentation, wasting hours at a time tinkering in PowerPoint when they should be researching their content and honing their script. The result is weak presentations with messages lost in the tacky blur of busy, multicolored slides that make funny noises and contain distracting animated images. Remember, your ultimate goal for visual aids is to support getting your main (oral) message across to your audience.

Slides

When writing your slides, apply the design principles used for writing resumes. Like the resume your slides are an **outline**. Don't use complete sentences, and avoid text-heavy content. Use concise lists in sentence fragment form. When choosing fonts, sans serif fonts like Helvetica or Arial are best. Choose no more than two font families. Use 24 point font and larger, nothing smaller than 18 point. Avoid the use of italics and all capital letters. Be sure to use **contrasting colors** between the text and background for readability. Do a "back of the room" test to see if there are any problems with your design choices.

Include the following elements in your slideshow:

- **Title slide:** The first slide captures the main point of your presentation and contributes to the initial impression you make with your audience. Try to avoid generic titles. Instead of "PowerPoint," how about "PowerPoint Exposed: Revealing the Secrets of Designing with PowerPoint 2000"? Include other details like your name, the date, and possibly the setting of your talk. Add specific details to your slide to let your audience know you tailored it to them (e.g., the correct date or the company's logo pulled from its Web site).

Figure 8.1. *Example title slide*

- **Overview Slide**: The second or third slide should be a bulleted list of the main points of your discussion.

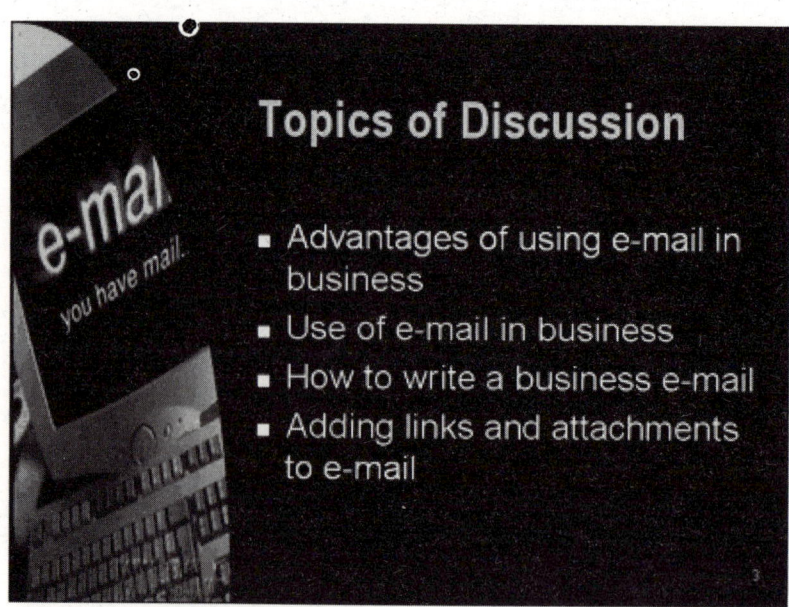

Figure 8.2. *Example overview slide*

- **Body Slides**: Format your body slides with consistent use of design elements like font choice, font size, color scheme, location of text, bullets, etc. Use the principle of "talking headings" for each body slide. In other words, give each slide a heading that captures in a short phrase the gist of that particular slide. If you have too many points to fit onto one slide, create a second slide with "title (con't.)."

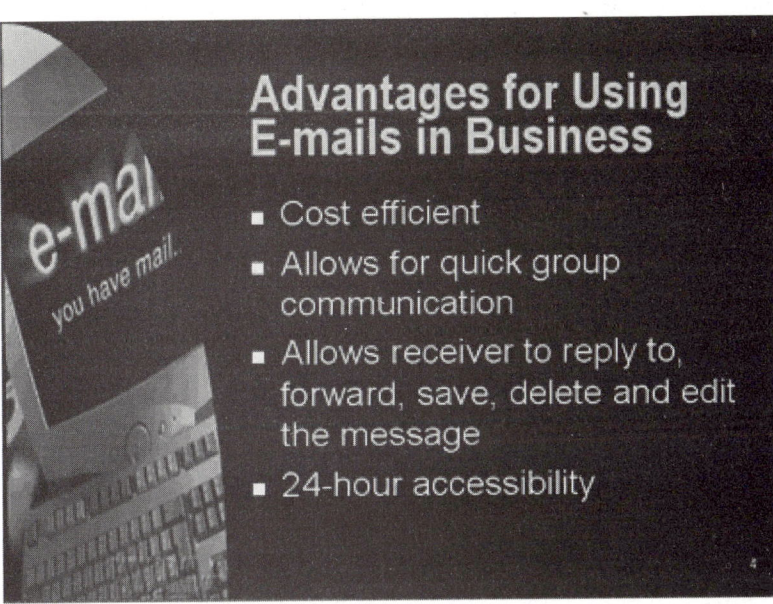

Figure 8.3. *Example body slide*

- **Closing slide**: End by reiterating your main points or outcomes. Afterward, you can consider a clever and/or humorous final slide that leads into the Q&A period, but avoid clichéd endings like a big "The End."

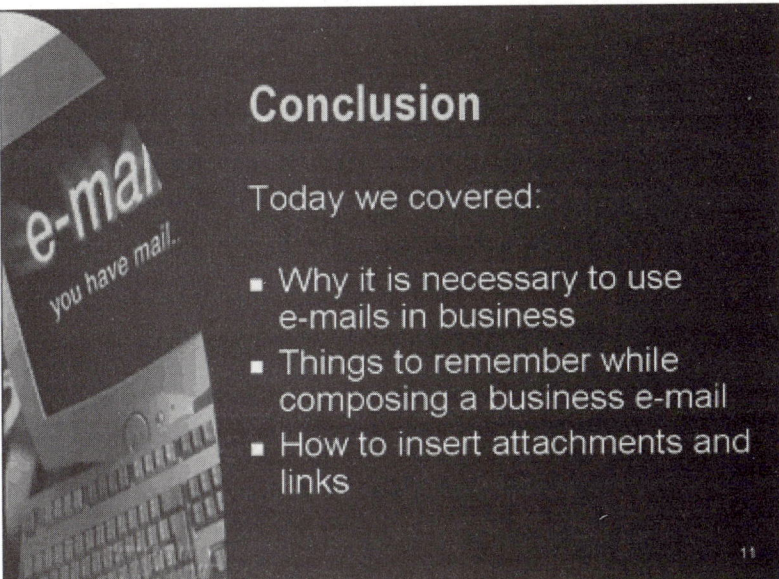

Figure 8.4. *Example closing slide*

Chapter 8: *Presentations*

Other important general guidelines when designing slides include:

- As a rule of thumb, plan to **use no more than one visual aid per minute**; a fifteen-minute presentation will require no more than fifteen slides, and preferably less. Don't overload your audience with specifics they will never remember.
- Templates are helpful for ensuring design consistency, but consider strongly the implications of choosing a template or using a template wizard when producing slides or transparencies. Does the template best represent the image/message you want to convey? Is it appropriate? Is it easy to read? Is it too busy with color and graphics? Very often, novice presenters fail to fully consider the implications of the template they choose. Don't pick the tropical beach-themed template because you want a vacation, but because you're presenting to a group of travel agents.
- Pay attention to the design principle of balance. Your combination of text, graphics, and white space should not appear unbalanced. If items in a bulleted list appear to take up only the top or left portion of your slide, add a graphic (a photo or clipart image) to balance the layout and provide visual appeal. When choosing graphics avoid picking amateurish looking clip art and loading a single slide with too many graphics. Another element of balance is to aim for a consistently sized border of white space around all sides of each slide.
- If you cite statistics, quote someone, or borrow a graphic, be sure to identify your source. In a small font somewhere at the bottom of the slide or near the graphic, write, for example, "Source: U.S. Dept. of Labor Quarterly Report, Oct. 9, 1999." Be prepared to provide the full citation should someone ask, or include it in a handout.
- Move dense, text-heavy content to handouts. Put long quotes, text-heavy tables, or complex figures on paper instead of creating a hard to read slide that will be more distracting than helpful. Use concise slides in conjunction with the printed handouts that contain the dense content.

Handouts

These are among the easiest visual aids to make. They can also be a valuable tool for keeping your listeners on track during the presentation, ensuring that they go away with accurate information, and providing them with permanent reference material. Handouts don't have to be elaborately made, but the more professional looking your handout, the better the impression you make with your audience. As with any document, you should always consider issues of page design as an element of the overall usability and persuasiveness of whatever handouts you distribute.

Should you distribute handouts before, during, or after your presentation? It depends. You should time the distribution of your handouts to coincide with the presentation's purpose. If your purpose is to inform your audience, you can distribute exercises and workshop materials before the presentation. But an article, traditional formal report, or other reference information probably should be distributed at the end (unless you're planning on referencing or explicitly referring to such information during your talk). This keeps your listeners from becoming readers until *after* the presentation. Sometimes you can coordinate handouts with other visual aids. For example, a procedures manual or a list of features and benefits might be distributed as a model when introduced.

As with every form of visual aid, the most effective handouts are those that are actively incorporated into your presentation. Introduce the purpose of your handouts and tell your audience how to use them. Consider using a fill-in-the-blanks approach: get your listeners actively involved by making them follow along with the handouts as you speak. One literal interpretation of this is to have them fill in information or answer questions on the handouts as you speak.

Always bring enough handouts for everybody plus a few extra. A good rule of thumb is to use the room capacity of your presentation or to find out ahead of time the estimated or real attendance figures.

Delivery

Your goal is to make yourself, your presentation, and your key points as memorable as possible. Preparation is the key. Know your purpose and audience. Prepare your content so the number of key points is limited, so they are presented in a dynamic and convincing way, and so they are repeated strategically (so they are signposted in the introduction and conclusion).

Develop a **script**, a detailed outline that includes directorial cues like when to change slides, when to pass out flyers, and who's going to say and do what (see Figure 8.5). Most scripts are written verbatim, but with the knowledge that they will not be read verbatim. The more details you write into your script, the more likely you are to remember them as you review and rehearse. Develop your script based on the total time you have to present, and test and refine your timed outline with actual practice. This will give you greater control later, when you are presenting from your notes or slides.

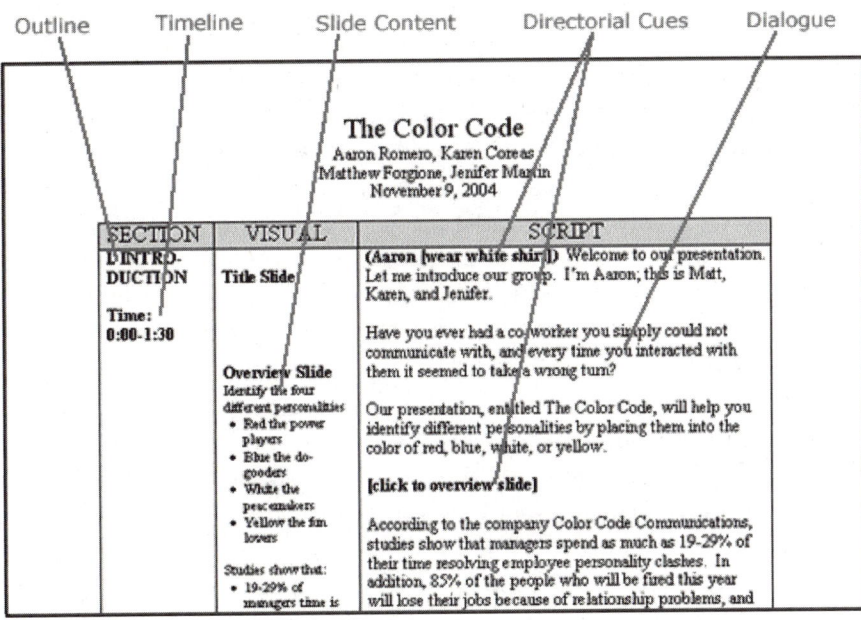

Figure 8.5. *A good script contains more than just an outline*

The other essential ingredient is to prepare yourself by **rehearsing**. The two deadly sins of presenting are being unprepared and speaking for too long. These sins go hand-in-hand. If you are prepared, you will have rehearsed, and if you rehearse, it is easy to time your presentation. You cannot omit this essential element in preparing your presentation.

Once you've written a complete script, practice several times from it. For high-stakes presentations experts recommend a minimum of five trial runs—the last at least two days before the actual scheduled event. You can adjust your practice time accordingly, but you should never walk into an important presentation never having practiced. Stand before a mirror, a video camera, or better yet, an impartial group of observers. Watch and listen to yourself. Play and replay the recordings. Get feedback from the trial-run observers. Ask yourself and your observers whether the main points are clear and receive adequate emphasis throughout your talk. Time the presentation to make certain it isn't too long. Repetition will help you incorporate the specifics of what you want to say and the rhythms with which you want to deliver them into your subconscious so that the actual presentation will be second nature to you.

Other points to keep in mind when preparing and practicing your delivery include:

- **Avoid stuttering** and using too many "ahs" and "ums" and "like, you knows." If you have a tendency to talk like this you need to practice them out of your presentation, and consider taking a speech class to polish your delivery techniques.
- **Don't read from each visual verbatim**. Your audience members can read at their own pace (often faster than yours) when it's convenient to them. Make your role as presenter indispensable: explain, embellish, and add detail. Elaborate on each visual with your discussion. Talk to your audience, don't read to them. Don't expect the slides to do the work for you.
- **Never turn your back to your audience**. Instead, print notes or slide views from your PowerPoint file to keep your slide content in front of you while your actual show is projected onto a screen behind you. If you have to point at something in the slides, keep your toes and shoulders pointed at the audience as you reach back. Avoid using laser-pointer devices. Once a very popular presenter's toy, they're now considered amateurish and distracting.
- **Always try to involve your audience** with eye-contact, questions, and interactive elements within your presentation. While this depends on the context of your presentation, typically audience members are more likely to enjoy and remember your presentation if you involve them as more than passive receptors of your discussion.
- **Include checkpoints** if you're leading an activity or demonstrating how to do something. Stop and check to see if anyone is lost or needs help. Have assistants circulate to assist individuals who fall behind or get stuck.
- **Always get there early** and make sure the equipment/technology is present, working, and suited to the site. Start on time and end on time. Appoint a timekeeper before you begin. If your session is 12 minutes, talk for only 10 and leave time for questions and answers. Dress appropriately

and professionally. If you're unsure what's appropriate for a particular context, err on the conservative side.

- **Be yourself when answering questions**. Questions are an opportunity to elaborate on your message around an interested audience member's pointed question. Take advantage by saying more than yes or no. If you don't know the answer, don't be afraid to admit it (but don't get caught not knowing something central to your purpose). If you don't understand the question, ask the person to rephrase it. If you should happen to have someone put an angry or hostile question to you, maintain your composure. Politely and respectfully try to respond; avoid getting into a figural—or literal—boxing match. Maintain goodwill.

- **Prepare a back-up plan for high-stakes presentations.** How will your spectacular PowerPoint presentation fare if there's no computer onsite? What if it dies right in front of you? Always prepare a back-up set of visuals in another medium. Make handout copies of your slides or bring your slides on Posters, Flip-Charts, or Easel Board. Of course cost is a factor, but when losing a client or professional ethos within your own company is a factor isn't the extra cost worth it?

- **Keep vital presentation elements safely with you if you** are traveling, not in your luggage bag that may wind up in Tahiti. Murphy's law, "anything that can go wrong, will go wrong," applies doubly to presentations and, particularly, computer-assisted presentations.

The Big Day

Delivering high stakes presentations is never easy. It will always require planning and you'll always be worried about making the deal or accomplishing your goal. But it gets easier with each experience presenting. The best way to minimize nervousness is to prepare and practice. If you know what you're going to say and have practiced how you're going to say it, you'll have less to worry about. If you're worried about being the center of attention, know that a well-prepared set of visual aids will help divide the audience's attention between you and the visuals. There's nothing wrong with this, as long as your visuals are professionally designed and contribute to your purpose.

Lastly, don't worry about making mistakes or slip-ups during the presentation. If you slip, just keep going; it's likely your audience won't even notice. If it's a bigger slip, that the audience does seem to notice, make a joke and then go on. Your audience will appreciate your composure.

Online Resources

- *"Developing a Presentation in Four Easy Steps" by Presenters Online*—Concise overview of presentation planning considerations with links to additional information on key points. http://www.presentersonline.com/basics/delivery/4easysteps.shtml
- Presenters Online.com— This site offers advice from presentation consultants, plenty of tips and tricks, and a bank of great clip art to enhance the creation of slides. http://www.presentersonline.com

Chapter 9

Collaboration

Teams improve communication by sharing information within an organization. Because of their versatility teams can produce large amounts of work, including written documents, that individuals would have difficulty accomplishing on their own. Teams can increase productivity when writing collaboratively by brainstorming ideas, dividing responsibilities, working together to gather information, and editing each other's work. Effective teams can solve problems creatively and efficiently. The better equipped you are to analyze what is happening in and around your team, the more successful and satisfying your group writing project will be.

Developing Team Skills

During the first stage of team development members of the group need to get to know each other. Trading contact information and discussing the skills that each member feels comfortable contributing to the group, as well as being honest about fears and weaknesses, will help individual members know what to expect from each other.

Conflict

As teams develop they should expect conflict to arise. How a team manages conflict determines the quality of its performance. Teams may experience two kinds of conflict: cognitive conflict and affective conflict. Cognitive conflict revolves around project-related issues and is a necessary and healthy aspect of a well-functioning group. Creative thinking and stimulating discussion should be the result of cognitive conflict. Affective conflict centers around feelings and personalities rather than issues—on people rather than the matter at hand. Affective conflict tends to be emotional and may erupt into personal criticism, which is destructive to trust, a necessary component for the team's success. *Groupthink*, a term coined by Irving Janis, may be equally destructive. Groupthink occurs when team members are overly eager to agree with each other and fall victim to faulty decision making processes. Teams suffering from groupthink may fail to examine all alternatives, self-censor thoughts and ideas, and refuse to question the decisions of a strong leader. Effective teams encourage open discussion, focus discussion on issues, and evaluate many alternatives carefully.

Communication

Although the end goal of the group project may be a written document or documents, clear verbal communication is also necessary. The best teams speak clearly, contribute ideas freely, and encourage feedback from one another. Constructive disagreement is encouraged. Teams whose members enjoy working together take their tasks seriously but are also able to laugh at themselves and inject humor into their interactions. Two of the most important tools for effective communication are the abilities to listen and to read body language.

Even though communication in groups is often informal, conscious efforts to listen will increase the team's ability to work well together. To increase listening skills:

- Stop talking and focus internally on the speaker (block out competing thoughts)
- Maintain an open mind
- Don't interrupt; lean forward and maintain eye contact with the speaker
- Don't fidget or try to complete other tasks while listening
- Ask clarifying questions
- Rephrase and summarize what the speaker has said in your own words; ask the speaker if your summary is a correct interpretation of what he/she intended to communicate
- Listen between the lines (observe nonverbal cues)

Nonverbal communication, or body language, often tells the listener more than the speaker's words. Just as you use nonverbal language to communicate your interest in someone else's words (by leaning forward, making eye contact), it is important to pay attention to the signals others are giving you. Leaning forward, opening hands, sitting on the edge of the chair, placing hands behind one's back, and sitting in a relaxed position show cooperation and feelings of confidence, while fidgeting, arm crossing, eye rubbing, and throat clearing may indicate defensiveness or nervousness. Hunching over, pen chewing, hand wringing, and neck rubbing are signs of insecurity or frustration. However, as people are individuals, they each may have their own ways of communicating different emotions. The goal is to pay attention to the clues that the speaker is giving, nonverbally, and if you detect cues that seem to contradict the speaker's words, politely seek additional clues by asking questions:

- "Please tell me more about…"
- "Do you mean that…"
- "I'm not sure I understand…"

Asking questions of this nature reassures the speaker that you are interested in understanding his/her real feelings and message, which will go a long way toward establishing trust and open communication.

Compromise

When team members cannot agree, negotiation and compromise may be necessary. It is important to remember, however, that when a successful negotiation occurs both parties win. During negotiation listen carefully, be empathetic, explain your position logically and clearly, avoid expressing hostility or engaging in personal attacks, be willing to compromise and expect compromise from others, and be willing to follow through with whatever you have agreed on.

Team Leadership and Roles

Unless your team has been assigned a leader, you will have to organize your group on your own. Effective teams utilize the idea of roles within the group to maintain order and produce work efficiently. Sometimes a leader emerges naturally and members of the group volunteer for different positions. Positions that are important in all teams, whether the end product is a written or oral report, include:

- Leader (to plan and conduct meetings)
- Recorder (to keep a record of group decisions)
- Evaluator (to determine whether the group is on target and is meeting its goals)

Teams also must decide whether they will be governed by consensus, where everyone must agree, or by majority rule.

The role of the leader should be to facilitate meetings—not to do all the talking or make all the decisions. The purpose of meetings is to exchange views, and the leader's role is to encourage all members to give their opinions, ideas, and responses. The leader can also encourage the group to avoid digressions and to adhere to a schedule. Finally, when the group seems to have reached a decision, the leader can summarize the group's position and check to see whether everyone understands and agrees. No one should leave a meeting without full understanding of what was accomplished. An effective leader may ask each person to recap briefly the content of the meeting and decisions that have been reached in order to clarify that each person understands.

An important item to discuss when a team first begins to meet is how to deal with group members who fail to fulfill their duties or pull their share of the load. The leader of the team, along with the evaluator, can work together to keep members of the group on task. Including a brief report (along with the final written report) which discusses how each member of the group contributed to the overall project will be incentive for individual members to fulfill their assigned responsibilities. Or, a chart may be used in which each member of the team rates the other members from 1 to 5 on how they performed in the following areas:

- Contributed his or her fair share to all phases of the project
- Participated actively in meetings

- Was dependable, prompt, and courteous as a group member
- Overall rating for this person's contributions to the group

Organizing and Managing the Project

The most effective way to organize a project is to make a project plan. The team begins by discussing how to break the project into parts and assign responsibilities to each member, how to collect the most accurate and useful information, and who will be responsible for gathering what information. Once the project has been divided into discrete parts, the team establishes deadlines for completing tasks and for writing and editing the document.

The complete project plan should involve a list of the **tasks** that must be accomplished, **who** will be responsible for completing them, and a **schedule** or timeline for the process, including intermediate and final deadlines. A good place to begin is to create a PERT or Gantt chart for the project. PERT charts, short for Program Evaluation and Review Technique, map out various paths of work for the project and include discrete project phase start dates and end dates.

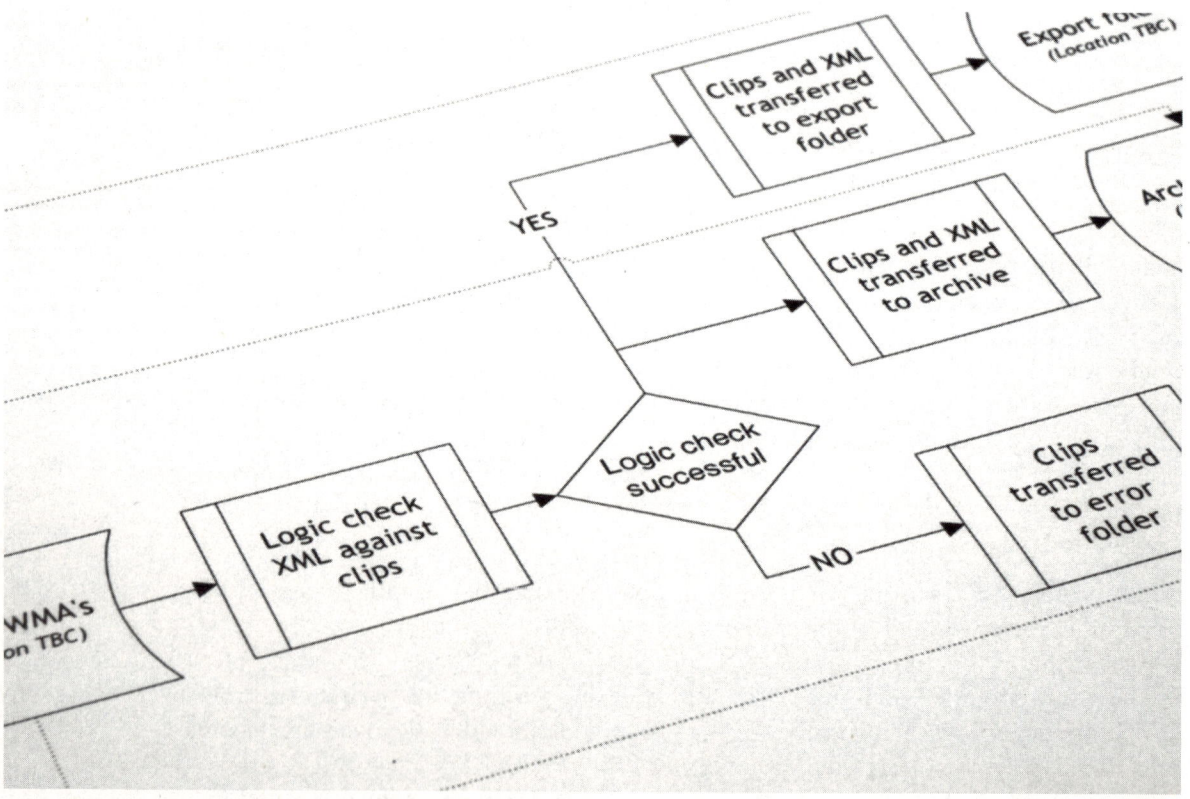

PERT chart

Gantt charts show various overlapping tasks in a project as well as indicating the duration of each task. Using a Gantt chart enables team members to see where they are in the process, who has successfully completed tasks, how much time is left to complete those tasks remaining, and where adjustments need to be made in time allotments.

ME 5254 Canoe Lift Project, SQ 97
Updated Apr-7-97

Task	Duration	Who	APRIL 1	3	8	10	15	17	22	24	29	MAY 1	6	8	13	15	20	22	27	29	JUNE 3	5	9	11
MARKET RESEARCH																								
Interview 25 customers	2w	dz,mp,sq			■	■	■	■																
Internet survey	2w	kl			■	■	■	■																
COMPETITION																								
Brochures/ads/catalogs	12d	wd,kl		■	■																			
Sporting stores	12d	wd,sq		■	■																			
OTHER RESEARCH																								
Patent search	2w	mp,dz		■	■	■	■																	
CONCEPT DESIGN	7d								■	■														
Brainstorming session(s)		ALL																						
CONCEPT SELECTION	2d	wd								■														
DETAIL DESIGN																								
Structural analysis	14d	sq										■	■	■	■									
Pro/E drawings	14d	kl,mp										■	■	■	■									
Mechanism design	14d	mp										■	■	■	■									
PROTOTYPE																								
Order purchased parts	7d	kl												■	■									
Fabricate machined parts	14d	dz												■	■	■	■							
Assemble	5d	dz,kl,mp															■	■						
Test	5d	dz,kl,mp																■	■					
Alpha prototype complete	0d																		•					
Customer reaction tests	5d	sq																	■	■				
Refine, build beta version	7d	dz,kl,mp																	■	■				
MQ REVIEW																								
Plan	12d	kl							■	■	■	■												
Create overheads	2d										■													
Rehearse presentation	1d											■												
Modify overheads	2d											■												
MQ Review	0d											•												
DESIGN SHOW																								
Plan	2w	dz															■	■	■	■				
Fabricate exhibit	5d																			■	■			
Design Show!	0d																					•		
REPORT																								
Plan	2w	mp																■	■	■	■			
Write draft sections	9d	ALL																		■	■	■		
Edit draft	2d	mp																				■		
Produce final	2d	mp,sq																					■	
Report due!	0d																							•

Gantt chart

> Durfee and Chase's "Brief tutorial on Gantt charts" is a very good quick guide to creating a Gannt chart. URL: http://www.me.umn.edu/courses/me4054/assignments/gantt.html

Projects don't always go exactly as they are planned; however, a project plan enables all members to see how things are progressing and to make changes when necessary. Having a schedule and deadlines for a project is essential to a team's ability to complete the project successfully.

Document Production

The advantages of teamwork can be easily sabotaged without careful consideration about how to maximize the range of ideas and strengths that members bring to a group project. After developing team skills and planning the broad framework of your project the labor intensive tasks of drafting, revising, and editing present new challenges.

While the principle of writing collaboratively may seem simple, many find that the actual process of collaborating is contrary to the majority of the educational experience, which requires one to function as an individual author. To reap the benefits of teamwork it is vital that your group consider exactly *how* it will write together.

Plan Early

Half of the battle of creating a successful group project is a commitment to leave nothing to chance. One way to ensure members are willing to make such a commitment is to involve *all* group members in the initial planning, regardless of the role they will play in the project (leader, recorder, evaluator, etc.).

In addition to creating a project plan sets up a schedule for key deadlines and the division of project tasks, it is important to establish frequent group meetings to serve as checkpoints throughout the project. One of the most frustrating pitfalls of writing collaboratively is for one group member to invest a substantial amount of time and energy on a project task that is not essential or becomes obsolete. Frequent checkpoints allow the group to continually assess its progress and refine or alter goals if necessary.

Three helpful strategies beyond project scheduling for maximizing your group's document production time are:

1. Use a style guide
2. Create a storyboard
3. Use technology

Use a Style Guide

One of the most difficult challenges in any collaborative writing project is maintaining consistency throughout the document, especially when parts written by various team members are brought together into a master document. Groups can reduce their workload later in the project by using a style guide to help ensure individual writers are using a uniform style from the beginning of the drafting process.

Style guides dictate document design aspects such as the formatting of page layout, headings, body text, and graphics. Style guides also set up guidelines on mechanics such as what kind of abbreviation to use, whether to spell out numbers, what rules for comma usage to follow, and how to treat nonstandard spellings such as "e-mail" vs. "email" or "Web site" vs. "website."

General style guides such as the *Chicago Manual of Style*, the *CBE Guide* (for writers in the biological sciences), or the *APA Style Manual* (for writers in the social sciences) are all good choices. However, your group may want to sit down at an early meeting and create its own particular style guide. Certainly it will be helpful to make early decisions about how to handle the jargon or industry specific language that may surround your project topic.

Create a Storyboard

Based on the way Hollywood screenwriters depict a movie's story prior to filming, storyboarding as a practice of document production enables team writers to draft separate portions of a document so that they fit together seamlessly and require a minimum amount of revision. The storyboarding technique works like this: after creating a comprehensive document outline in the planning phase of the project, the group assigns sections to individual members. Before the next meeting each member creates a storyboard for his/her section(s). Typical storyboards address the following:

1. Summarize the particular document section in one complete sentence.
2. List the key points that will be developed in the document section.
3. Note if a visual should be included, and, if so, the purpose, type, and title of the graphic aid.

When the group meets again, members put all the separate storyboards together. Groups often tape their storyboards on a wall or make photocopies of all the separate storyboards prior to the meeting. Bringing the storyboards together enables the group to form a comprehensive vision of the final document and to see how each section works together. This process often reveals overlapping information, inconsistencies, and omissions, but most importantly, after team revision, your group's storyboard serves as a detailed guide for drafting!

Use Technology

Because collaborative writing projects involve individual work as well as group work, it is vital that groups communicate regularly. Technology offers business writers effective and easy ways to enhance collaborative writing tasks.

- **E-mail**: E-mail is a perfect vehicle for keeping all group members abreast of progress, planning or changing meetings, asking each other questions, transmitting drafts back and forth, etc. Although e-mail is by nature a more casual form of communication, it is important that all group members maintain their professionalism.

- **Word Processing**: Microsoft Word offers several tools for drafting and revising documents. The most useful collaborative tool is the **comment** function. Selecting the **View>Toolbars>Reviewing** tool enables writers and reviewers to add comments to a document—small post-it style notes or editor's remarks. To do this highlight the text you wish to comment on, select **Insert>Comment** (there is also a yellow, post-it-like "comment" button), and then type your comment (see Figure 9.1). The comments appear in different formats depending on what version of MS Word you're using.

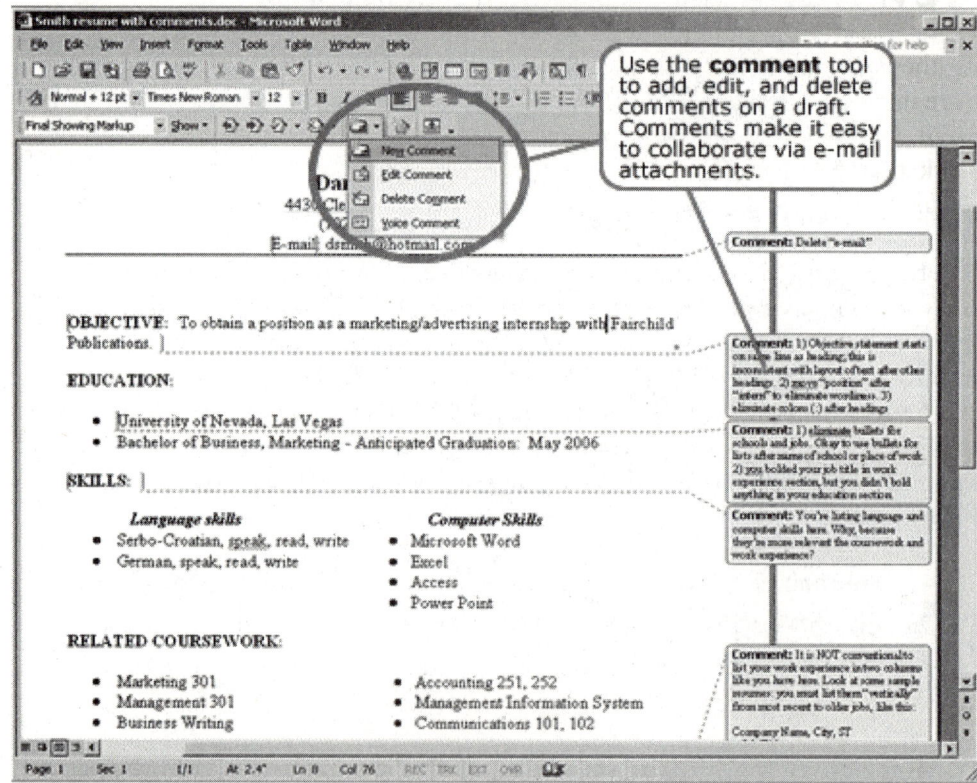

Figure 9.1. *The "Comment" tool makes it easy to collaborate via e-mail*

Other collaboration aids include the **track changes** tool, which creates a record of all changes made to a document. While this feature is a bit cumbersome for basic student group writing tasks, some organizations require it. The **highlighter** tool can be used to mark typos or color-code types of mistakes.

Don't overlook basic file-management issues when working collaboratively. Files should be named in such a way as to be easily recognizable and prevent accidental overwrites. Successive versions are usually numbered (e.g., report_draft_1, report_draft_2, etc.). Everyone should keep backup files of all document versions in case any one person should lose a file by accident or computer malfunction.

- **Computer Conferencing**: Business meetings are increasingly held online using collaborative software called groupware. Computer conferencing requires group members to log on to a specific server and use a software program such as Microsoft Netconferencing or commercial vendors like WebEx.com. This method of communication is particularly helpful to groups whose members are working in different cities. Other groupware allows members to work on the same document either at the same time, synchronous, or at different times, asynchronous. Groupware enables members to work almost as if they were sitting in the same room because an individual can view the project document in one window while making comments and changes on another.

Revision and Editing Stages

By the revision and editing phases of a collaborative project all group members have invested considerable effort into drafting their assigned sections of the document. Group members may feel a certain sense of ownership to their individual work that can create a barrier for effective revision. Paul Anderson, author of *Technical Communication*, offers two strategies for effective team writing:

1. **Be considerate when discussing drafts**: Acknowledge that group members have invested individual skill and creativity in producing their portion of the draft. One way to show consideration is to pose your suggestion as an option. You could say, "Here's another way you could communicate your point…" Another strategy is to focus on the positive reasons for choosing one option over another. As always, coupling suggestions for change with praise for strong features of a draft usually helps the writer member realize that the suggestions are not criticisms of that person's overall writing ability but specific instances where the writing is less effective. It is also important when your portion is up for critique that you set aside your ego and present yourself as open to constructive criticism. Oftentimes group members will fail to make important revision suggestions for fear of hurting a group member's feelings. In this situation no one wins.

2. **Treat drafts as team property**: In order to revise fully and adequately, group members must (1) let go of personal ownership in their portions of the draft and (2) take ownership of parts written by others. The document is the product of the entire team, thus everyone is responsible for ensuring that each section is effective and coherently fits with the rest of the document. This does not mean that individual group members should passively yield to any and all revision suggestions, but that all group members should engage in a reasoned discussion about the best way to write all portions of the *team's* communication. One way to decrease personal ownership is to alternate responsibility for portions of the document in subsequent drafts. For example, the team member who was responsible for writing the background section of the first draft would be responsible for the recommendations section in the second draft.

Revise and Edit Thoroughly

Groups are often lured away from effective revision and editing practices by the seduction of a completed first draft. Don't be duped. All effective, professional communication needs to undergo a thorough revision process. Ideally your group should solicit feedback after the first draft has been completed from (1) your instructor or supervisor, (2) peers or coworkers, and (3) your internal team members. Carefully consider the feedback and revision suggestions your group receives and decide how to revise together.

Revision applies to global considerations related to the rhetorical context of the document. Before worrying about checking for surface issues like grammar, punctuation, and spelling, the team should focus on big-picture document issues such as focus, key points, organization, and content. The group should check:

- **Plans vs. draft**: Does each section adhere to plans established in initial outlines and storyboards? Was anything unintentionally left out? Did new content work its way in, and, if so, is the new content relevant and useful?
- **Reader's point of view**: Does the document answer all the major questions readers would have such as purpose of the document, problem being addressed, research conducted to address the problem, presentation and interpretation of research findings, and cogency of conclusions and recommendations being made? Does information provided, claims made, or actions requested take into account the needs, values, and background of the reader?
- **Flow**: Does each section flow coherently into the next in a way that readers can easily follow? Clear headings, introductions/openings, transitional phrases, and topic sentences all serve as cues that help readers comprehend the structure of your document.

Editing, a microscopic investigation of the surface features your document should happen after all major revisions are complete. As the last stage in your collaborative project before submitting a final draft, editing is the way to catch seemingly minor errors that have escaped everyone's attention up until this point. Professional communication professor James Porter advises that editing should be done by more than one person in more than one setting. He suggests that editing tasks should be divided among group members in the following manner:

- **Document design**—Are there consistent margins, pagination, position of text, spacing, indentation, column alignment, layout of pages?
- **Names, titles, formalities**—Are all important names and nouns correctly spelled, capitalized, and with proper titles acknowledged?
- **Numbers**—Are numbers consistently written, using either words or numerals? Are numerical totals, data, and arithmetic correct?
- **Grammar, word usage, punctuation, and spelling**—Are all periods, commas, semi-colons, colons, and apostrophes in the right place? Has the computerized spell-checker missed any usage mistakes (e.g., too, two, to) or other mistakes (e.g. a singular word spelled correctly but that should be plural).

- **Visual aids**—Are all tables and figures consistently designed and labeled?
- **Macro view**—Are all parts of the document together and in order?

After all the hard work and time you have invested in creating the document, in the editing phase make certain that every aspect of the document is representative of the group's commitment to excellence. The beauty and perhaps the pressure of collaborative writing is that there is always more than your own individual professionalism on the line.

End Notes

Anderson, Paul V. (1999). *Technical Communication: A Reader-Centered Approach.* 4th ed. Fort Worth: Hartcourt Brace College Publishers.

Goodall, H. L. & Goodall, Sandra. (2002). *Communicating in Professional Contexts.*

Guffey, Mary Ellen. (2000). *Business Communication: Process & Product.*

Hamilton, Cheryl with Parker, Cordell. Communicating for Results: A Guide for Business and the Professions.

Jones, Dan and Karen Lane. (2002). *Technical Communication: Strategies for College and the Workplace.* New York: Longman.

Lopez, Elizabeth S. (1994). "Collaboration and Project Management." Purdue Business Writing Coursepack. Purdue English Department, Lafayette, IN.

Markel, Mike. (2001). *Technical Communication.* 6th ed. Boston: Bedford/St. Martin's.

Pfeiffer, William Sanborn. (2003). *Technical Writing: A Practical Approach.* 5th ed. Upper Saddle River: Prentice Hall.

Porter, James. (1994). "Strategies for Team Drafting and Revising the BCIA Report." Purdue Business Writing Coursepack. Purdue English Department, Lafayette, IN.

"Collaboration" contributed by Constance Pruss and Jenny Toups

Chapter 10
Definitions

Technical and business communicators are often asked to define concepts, terms, and processes for audiences who have little or no understanding of a particular field's jargon or a company's product, system, or process. Technical and business writers are also frequently asked to provide precise definitions to explain complex legal, social, ethical, and moral issues.

Knowing the level of technicality of an audience (non-technical, semi-technical, or highly technical) helps writers determine which concepts within a document may require definition. It may be helpful to view some document types, such as an instruction manual, legally binding policy, or even a business prospectus, as an instance where definition is one of the underlying purposes of the document. The strategies for writing definitions discussed below can help writers in any situation when the task at hand requires explaining complex ideas clearly.

Types of Definitions

When explaining a complex idea to a reader unfamiliar with the concept, there are three types of definitions at the writer's disposal:

- Parenthetical definitions
- Sentence definitions
- Expanded definitions

Parenthetical Definition

A parenthetical definition is a word or brief clarifying term placed within a sentence. These are not intended to be comprehensive but are mainly meant as a short definition of an unfamiliar term.

Example:

The aircraft accident report concluded that the crash was the result of faulty maintenance on the right aileron (moveable wing flap).

Sentence Definition

At times you will need to write a more detailed definition for a complex or unusual concept. The sentence definition, which may be more than one sentence, is based on a pattern of composition that seeks to include as much information as possible in a minimum amount of space. Sentence definitions are composed of three parts:

- The **term**, either a word or a phrase, to be defined
- The **class** to which the term belongs
- The **distinguishing characteristics** that make it different from all other terms in its class

When developing a sentence definition use the following pattern:

Term	Class	Distinguishing Feature
Angina pectoris	heart condition	severe chest pain caused by insufficient blood supply to the heart.

These three elements are then combined to form one or more sentences:

Example:

Angina pectoris is a heart condition caused by an insufficient blood supply to the heart resulting in severe chest pain.

Expanded Definitions

There are times when parenthetical or sentence definitions are not satisfactory. This more often than not occurs when defining vague or abstract terms. Terms such as *rental agreement* or *auto lease* can have a number of meanings and interpretation depending on the knowledge and previous experience of the writer and the intended audience. Therefore these terms and others like them need to have a clearly stated definition so that the writer and the reader have an understanding and a consensus on what terms mean and how they are used.

An expanded definition may be a paragraph, a multi-page document, or even an entire manual. Here is an example of an expanded definition of *deductible* from the Glossary of Insurance Terms on the Insurance Information Institutes Web site:

Example:

"The amount of loss paid by the policyholder. Either a specified dollar amount, a percentage of the claim amount, or a specified amount of time that must elapse before benefits are paid. The bigger the deductible, the lower the premium charged for the same coverage." (2005)

When composing an expanded definition, you will typically start with a sentence definition of the term and then move to one or more of the following strategies to develop a useful definition.

- **Etymology**: This method of expansion uses the origin of the word to help the reader understand the meaning.
- **History of the Term**: The meaning of a specialized word or phrase can often be defined by providing the reader with a discussion of how the term developed.

- **Compare/Contrast**: You may compare and contrast the term to information that the reader already understands.
- **Negation**: At times you may expand a definition by telling the reader what it is not.
- **Visuals:** Providing clearly labeled visuals will help to expand a definition.
- **Example:** It is frequently useful to provide examples that are familiar to expand your definition.
- **Process Analysis**: For complex multipart definitions, providing the audience with a list of the parts that contain an accompanying description of how the parts work and interrelate is one method of expansion that many readers find useful.

Writing Effective Technical Definitions

As with any workplace writing, composing and publishing an effective technical definition requires planning. Before composing a technical definition you should be able to answer the following questions in detail:

1. Who is the audience for the document?
2. What level of technicality is appropriate for this audience? (non-technical, semi-technical, highly technical)
3. Given the audience's level of technicality, what concepts within a document require definition?
4. What type of definition is suitable for each concept? (parenthetical, sentence, expanded)
5. For sentence definitions, have you considered the concept's class and distinguishing feature?
6. For expanded definitions, what method(s) will be used to fully develop each concept, and why?
7. Are there any ethical considerations for any of the definitions?
8. Are there any potential legal consequences for any of the definitions?
9. Are there any design and format issues that need consideration?

Potential Problem Areas

When writing definitions there are some potential problem areas watch out for:

- Be careful not to write a **circular definition**. This happens when the writer uses the term as part of the definition. For example, a writer wants to define *sexual harassment* and begins the definition with, "Sexual harassment is a form of harassment…"

- Don't be **overly technical**. To be effective a technical definition should not use terms that are too technical for the intended audience. Understanding the level of technicality that the audience possesses will help you to determine the type of language that is appropriate for the purposes of the definition.
- Avoid the use of **broad, unfamiliar, and abstract terms**. One of the purposes of a technical definition is to be useful, to be understood by the audience. Using terms that are too broad, that are unfamiliar to the reader, or that are too abstract undermines the effectiveness of the definition.
- Do not use *is when* and *is where*. These adverb phrases do not work well as an introduction to a definition. A useful rule of thumb is to use nouns and noun phrases to define nouns, verbs to define verbs, and adjectives to define adjectives.

Useful links

This list of links will help you to better understand some of the techniques for writing definitions.

- The Writing Definitions handout at the Purdue OWL provides another approach to writing a sentence definition. URL: http://owl.english.purdue.edu/owl/resource/622/01/
- The Illinois Institute of Technology Center for the Study of Ethics in the Professions "Codes of Ethics Online" provides an index of various ethics codes and codes of conduct.
- How Stuff Works.com. One of the important things that you need to be aware of about this website is the way it "defines" and "illustrates" complex technological concepts using "everyday" language and illustrations. An excellent example of this is the "How Blu-ray Discs Work" by Stephanie Watson: URL:
  http://electronics.howstuffworks.com/blu-ray.htm
- Professor Merrill Whitburn of Rensselaer Polytechnic Institute (RPI) Definitions Techniques handout on the RPI Writing Center Handout page provides a brief overview of some of the more common techniques used to write definitions. URL:
  http://www.rpi.edu/web/writingcenter/definition.html
- Duncan Kent & Associates Ltd's Technical Communicators Resource site is "a comprehensive resource for writers of technical publications." URL:
  http://www.techcommunicators.com/techcomm/index.html

End Notes

Burnett, R. E. (2001). *Technical Communication* (5th ed.). Orlando FL: Harcourt College Publishers.

Insurance Information Institute. (*n.d*). Glossary of Insurance Terms. Retrieved January 12, 2005 from http://www.iii.org/static/site/tools/glossary_frset.htm

Lannon, J. M. (2003). *Technical Communication* (9th ed.). New York, NY: Longman.

Markel, M. (2004). *Technical Communication* (7th ed.). Boston: Bedford/St Martin's.

Vanalstyne, J. S., & Tritt, M.D. (2002). *Professional & Technical Writing Strategies: Communicating in Technology Science* (5th ed.). Upper Saddle River, NJ: Pearson Education, Inc.

"Definitions" contributed by Homer Simms

Chapter 11
Usability Testing

Usability generally refers to how well a tool or technology helps someone perform a task. Web site usability deals with effective user interface design. The Usability.gov site defines Web site user testing as:

> Usability testing encompasses a range of methods for identifying how users actually interact with a prototype or a complete site. In a typical approach, users—one at a time or two working together—use the Web site to perform tasks, while one or more people watch, listen, and take notes. (http://usability.gov/basics/index.html)

How are Usability Tests Conducted?

Usability experts generally agree on a six step process for usability testing:

1. Determine the goals of the study
2. Develop a profile of typical users
3. Write the user tasks
4. Conduct user tests
5. Evaluate the data
6. Recommend or implement changes

Step 1: Determine the Goals of the Study

Your evaluation should target common tasks that users are expected to perform on the Web site. Identify the purposes or aims of the site. What are the intended uses of the site? Why do people come to the site? What tasks might they attempt to perform? Break these uses down into primary and secondary tasks. For example, a travel Web site's primary function is airline booking, a secondary task is finding information about travel destinations.

Step 2: Determine the User Profile

The goal is to match the people you user test with the target user profile. Identify typical users' attributes: Who will come to the Web site? What is their age, gender, education level, etc.? What will be their knowledge level about the Web site and its purposes? What will be their knowledge level about the Internet in general? What other Web sites are related to this and will the users be familiar with them? If you're using classmates as your user group, justify how they fit with the user profile.

Step 3: Write the User Tests

User testing involves observing actual members of the target group using the Web site and interviewing or surveying them before, during, and after they use the site to gather their feedback. Most experts agree that a well designed test of only a small group, 5 to 10 people, reveals any major usability issues with a Web site (see Jakob Nielsen's "Why You only Need to Test With 5 Users" http://www.useit.com/alertbox/20000319.html)

Once you've brainstormed a list of the site's primary and secondary functions, translate them into a list of tasks or scenarios you will ask your users to perform during the tests. There are two basic types of user tasks:

1. **Directed tasks**: These are specific, close-ended tasks related to primary or secondary purposes. For example, to study the usability of an airline's Web site, you might ask the user: "Book a flight from Cleveland to Phoenix for your Christmas vacation. You want to leave the morning of December 20 and return anytime on December 30."

> **Online Example:**
> - Jakob Nielsen's 1994 Usability Report, which used all directed tasks.
>   http://www.useit.com/papers/1994_web_usability_report.html

2. **Undirected tasks**: These are good for general evaluations/assessments (a good choice if you notice lots of problems with a Web site). Undirected tasks are more open-ended. They are scenarios designed to mirror how intended users might typically use the site such as: "You're at home browsing the Internet, interested in finding the latest news about Iraq. You decide to use the latest search engine being promoted by your ISP." Once the user is finished with the task you can collect their opinion about general and specific issues.

> **Online Example:**
> - SystemConcepts' usability checklist
>   http://web.archive.org/web/20020214042844/http://www.system-concepts.com/visitors/healthcheck.html

After you have developed your user tasks/scenarios, decide on a method for collecting user feedback. Choose a written protocol (distributed to a user who writes his or her own feedback) or an interview protocol (with answers recorded by researcher), or a combination of both.

Make sure you plan on asking enough of the right questions before, during, and after your user tests:

- **Pre-Test Questions:** Used to collect basic demographic information and possibly information related to purposes of your Web site (e.g., if it is an auto manufacturer's Web site, ask users if they have ever visited such a site before). Other demographic questions used to verify the validity of the user group include age, gender, and computer experience. You could also ask questions such as: How often do you use the WWW? For what length of time do you typically use the WWW? For what purposes do you use the WWW? Have you ever used to the WWW to investigate products you wish to buy? Have you ever purchased anything over the WWW? Have you ever purchased anything you saw advertised on the WWW?

- **Test Questions:** Used to prompt users' reactions to specific tasks: What are you doing now? Why? What is the first thing you're attracted to? What is a particular feature, for instance, background color, like? NOTE: some of the post-test questions below might work better as in-test questions given during or immediately after specific tasks.

- **Post-Test Questions:** Prompt users' overall reactions to the test and Web site: What was your general impression of each page? What problems did you have using the page and why? What do you believe caused this problem? What, if anything, helped you find the information you were looking for? What, if anything, made finding the information more difficult? What information would you like to see on the page that was missing? Was the page easy to use? Rate the pages you saw—which was best? Give three reasons.

Lastly, consider using other methods for collecting data:

- **Observation.** You should record key user actions, e.g., how long it takes to complete a task, where the user clicks, how many clicks it takes the user to complete a task, and what the user's page-to-page navigation is.

> See Sim D'Hertefelt's "Observation Methods and Tips for Usability Testing" for more discussion of useful observation techniques: URL: http://web.archive.org/web/20021016094606/http:/www.interactionarchitect.com/knowledge/article19991212shd.htm

- **Think-Aloud.** You can record what users say while performing tasks and verbalizing decision making process involved in performing a certain directed or undirected task.

Step 4: Conduct the Tests

This is where you get to actually see how users perform your tasks. Keep Keith Instone's advice, watch and learn, in mind as you conduct your tests. That is, don't interfere with the users as they attempt to perform the tasks. Avoid giving verbal or visual cues that might influence the users' interaction with the Web site. Do not give them any background information or help them if they get stuck. That's one thing you're looking for, if they get stuck. Take notes—write down what they say, what they get stuck on, what links they use, how much time it takes to perform tasks, etc. Ask follow up questions about what they had trouble with, their overall opinion of the site, etc. Use the questions from the protocol you developed beforehand. At the end of the test you can demonstrate for users what they had trouble figuring out.

Step 5: Evaluate the Data

The most important skill of the usability consultant is the ability to recognize patterns. As researchers and usability analysts, it is your team's task to make sense of the results of your user testing. You should first compile your findings (one of the sections of your final report will be a findings section). You can group your findings by tasks and by answers to protocol questions.

> For more on what it takes to be a good usability tester, see Jakob Nielsen's "Becoming a Usability Professional." URL: http://useit.com/alertbox/20020722.html).

Once you compile the data, you'll want to look for significant patterns: what did users have the most difficulty with? How severe were their difficulties? How satisfied were they with aspects you were testing? Once you start to ask questions like this, then you can begin to brainstorm solutions or fixes (your final recommendations). As you start to discover patterns in the results of your user tests, consult Web site design principles for expert opinions on how to improve these elements.

> Keith Instone's "First User Test" article is an excellent discussion of the findings and recommendations of his evaluation of the Internet Travel Network Web site
> http://web.archive.org/web/20030219012940/http://www.webreview.com/1997/05_30/strategists/05_30_97_8.shtml

Step 6: Recommend Changes

Most usability consultants can implement recommendations firsthand. Your team will have to instead write your recommendations in a credible and persuasive report. Again, Keith Instone's "First User Test" article is an excellent example of how detailed your recommendations should be. Notice that Instone makes three recommendations, all of which are linked to more detailed discussion.

Sample User Tests

The following are samples of user test protocols (or plans for conducting user tests) used by:

- Melissa Cheung and Anuja Dharkar's protocol draft for testing their "Policy Maker" software
http://ldt.stanford.edu/~adharkar/adharkar/policymaker/appendix/protocol.html
- Saul Carliner's "Simple Usability Scenario" example
http://saulcarliner.home.att.net/id/usabilityscenario.htm
- Wendy W. Naughton's Usability Participant Questionnaire template (a pre-test form, available from STC Usability Special Interest Group's Toolkit Page
http://www.stcsig.org/usability/resources/toolkit/toolkit.html)

Student Samples

Under the Samples page for this project, you can find user test protocols written by students included as attachments to the design plans for the first two projects, the School of Architecture and CCSN projects.

PART II

Projects

Chapter 12

Introductory Memo Project

Project Objectives

- Practice the conventions of writing business correspondence
- Learn the business memo format
- Inform your instructor of your background
- Practice using computer technology to compose a document from a template

For your first assignment you must write a memo that informs your instructor of your current academic status and other features of your background. This will allow your instructor to (a) find out about you, (b) assess your ability to write and follow directions, and (c) help your instructor place you in collaborative groups and situations throughout this course. Additional aims of this assignment are to introduce you to the basics of business memos and get you used to working in the course's computer environment.

Steps for Completing this Assignment

First, read the "Correspondence" chapter in this textbook. Your introductory memo should demonstrate knowledge of the principles reviewed in this chapter.

Next, download a copy of the Microsoft Word memo template from the Kendall/Hunt companion website. You must use to complete this assignment. If you've never downloaded a file from the Internet and saved it to a hard or floppy disk, seek help from either your business writing instructor, a UNLV computer lab assistant, or a friend/family member.

http://www.khwebcom.com/unlveng403

Then, using the template and your knowledge of the required reading for this assignment, please write a formal business memorandum that provides your instructor with the following information:

- **Personal Information**: What is your address, phone, and e-mail? Who is your advisor and how can your business writing instructor contact him or her? (NOTE: If you are in the College of Business, find this information from the college's Web site.)
- **Career objective:** What is your major, and what are your career goals?
- **Computer experience**: What is your knowledge of and level of comfort/anxiety with computers?
- **Computer access**: What is your access to computers? Since English 403 uses an online textbook you'll need Internet access. What kind of computer(s) will you be using for this course? Be as specific as possible about the type, processing speed, and version of Internet browser. Where are those computers located (home, school lab, work)? If you have any concerns about your access to a computer during this semester, please inform your instructor so he or she can help resolve any problems you might have.
- **Work/internship experience and workplace writing experience**: What work experience do you have and have you done any writing for/at work?
- **Other writing courses**: What other writing courses have you taken (e.g., composition) and where did you take them? How did you do in your writing courses? How would rate yourself as a writer?

You can also include in your memo whatever you think helps achieve the two aims identified in this memo's opening. For instance, if you have any concerns or anxieties about this course or if you have certain goals/expectations, it would be worthwhile to express them.

Follow the Required Memo Format

Unless your instructor specifies a different format, your memo should be no more than 1.5 pages and should follow the format, or document specifications, exemplified here. Include the memo heading from the template and use only 11 point Arial font (for headings) and Times New Roman font (for body text), single spacing within paragraphs, double spacing between paragraphs. Include bolded headings to help guide the reader through the memo, and use bulleted lists where appropriate. Unless otherwise instructed, you should follow this format whenever your instructor asks for a writing assignment from you.

Write Using a Business Style

This is your instructor's first impression of you as a person, a student, a writer. Treat it like you would other high-stakes, first-impression contexts such as a job interview or sales meeting.

Write using complete sentences and use a business style: write clearly and concisely, avoid overly pretentious diction, and edit the memo for neatness and correctness. Try to use headings and bulleted lists to present information clearly and quickly (as this memo does).

If there are too many mistakes your instructor will simply ask you to revise the memo until it is appropriately and satisfactorily written, just as a supervisor would in the workplace. But the damage to your identity as a careful and conscientious student/writer will have already been done, just like any other blown first impressions).

Submission Format

Your instructor will tell you how he or she wants you to submit your completed assignment. Some instructors prefer to receive two print copies (one for the instructor's records and one to comment on and return to you). Other instructors teach the course entirely online and will ask you to submit only an electronic copy.

Chapter 13
Situation Analysis

> **Project Objectives**
>
> - Practice analyzing a business writing situation
> - Apply concepts of effective business communication to the analysis of a specific writing situation
> - Demonstrate awareness of context, audience, purpose, ethos, and document design

Business and technical writers need to be aware of the following situational factors when composing documents:

- Context
- Audience
- Purpose
- Writer
- Document Design

Use the following prompts to brainstorm important factors that influence how documents are written. Writers who reflect on their writing situation before writing are more successful and efficient than those who don't. As you become more familiar with the situation analysis exercise, it will take less time and energy to complete.

Consider *situation analysis* as an informal prewriting exercise. Focus on getting your thoughts recorded via the prompts. Worry more about content than grammar or mechanics. The list of questions is not prescriptive. You may use as many or as few of the prompts as seems appropriate.

> You can download a copy of the Situational Analysis Worksheet as a Word Template (*.dot) file and save it to disk (or other easy to access location).

Preliminary Considerations

List the following:

- Writer's name, title:
- Department, project number or name:
- Subject/Assignment:

Audience

1. **Who is the primary audience for this document?**

    Be specific. If you are writing for an individual within an organization provide a name and title. If the document is for a broader audience, who are they? What assumptions can you make about the audience?

2. **What level of technical knowledge or expertise about the subject will the primary reader(s) have?**

    Is the reader an unfamiliar novice or an informed expert in the subject that you are writing about? Will you need to write a highly technical document, a semi-technical document, or a non-technical document?

3. **What preconceptions will the primary audience(s) have?**

    Are there any preconceptions that you can infer from your knowledge of the primary audience? How will the preconceptions influence the reader(s)? Will they be resistant to the information contained in the text? Will they agree/disagree with the information provided?

4. **Are there any cultural considerations that the writer needs to consider?**

    The WWW and advances in communication technologies have made the business environment in the new millennium global. How likely is it that someone from another country will read the document? Do you need to tailor the content to accommodate a worldwide audience? Do you need to at least acknowledge that this document will be read by an international audience?

5. **Who else will read the document?**

    You must assume that others, besides the primary reader(s) will have access to the document. Consider your supplementary or secondary audience. Who might that be, and why would they be reading the document as a secondary reader? For example, the company's lawyer may need to review your response to a client.

6. **What is the supplementary reader's level of knowledge or expertise?**

    Is the supplementary reader an unfamiliar novice or an informed expert in the subject that you are writing about? Will you need to write a highly technical document, a semi-technical document, or a non-technical document?

7. **What assumptions or preconceptions will supplementary readers have?**

    Are there any preconceptions that you can infer from your knowledge of the supplementary audience? How will the preconceptions influence the reader(s)? Will they be resistant to the information contained in the text? Will they agree/disagree with the information provided?

Purpose and Intended Use(s)

1. **What is the main purpose of this document?**

    Do you wish to convey information? Are you seeking to inform the reader(s)? Instruct them in a policy, method of operation, and/or a procedure? Are you offering a solution to a problem or concern?

2. **What are the secondary, tertiary, etc. purpose(s) of this document?**

    You may identify and explain any other purpose(s) that you have or that you wish to accomplish with this document. What else do you hope to achieve with this text? Don't forget about the important purposes of maintaining positive relations and creating a legally binding record.

3. **What information and/or content does the audience expect to find in the document?**

    Another way of approaching this question is to consider what the primary reader(s) are expecting to find in the document. Have you provided the necessary information and/or content required by the audience?

4. **How will the audience use the document?**

    What are the identifiable uses of the contents of the document? Having an understanding of how the audience will use the document will help you to compose an effective (i.e., useful to the reader) document. Will users rely on the information to build something, buy something, act differently, avoid harm or death, etc.?

5. **What is the best way to organize the information in the document?**

    Given the purpose and audience's intended use, how should you organize the document? Should your organization be direct or indirect? You could sketch a preliminary outline here.

The Writer

1. **What is your relationship to the primary reader(s)?**

    Defining and expressing the relationship between writer and reader(s) is important since this will influence much of the information in a document. Is the primary audience your teacher, manager, employee, client, etc.? What are you to your audience: superior, subordinate, friend, enemy, etc.?

2. **Given your audience and purpose, what tone should you adopt to convey your message?**

   In general, your tone should always be professional and sincere, but other moods may be appropriate such as humility to admit a mistake or courtesy to welcome a new employee, etc. You should generally avoid hostile or arrogant tones.

3. **Given your audience and purpose, what level of vocabulary should you employ?**

   Unless you are on a first-name speaking basis with your reader, assume a more formal style of language use. Should you avoid jargon, slang, or colloquialisms (conversational style)? Should you use first-person "I" and second-person "you" or should your voice be in the more traditional third-person "they" and "it"? Are there any concepts or ideas in your message that may have to be defined or clarified depending on your audience's technical background knowledge? For each term that needs clarification, what level of definition should be used?

4. **Are there any political or ethical considerations?**

   Do you risk stepping on anyone's toes with your message? For instance, how will you report a costly error to your company during your first assignment at work? Will your message impact others negatively or positively? Are there any financial, legal, or ethical factors related to your message that you should consider? How should you adjust your message?

Document Design

1. **What are the specific design components that are required to effectively complete the document?**

   This focuses on the issues of the use of a required template or format, type style, font size, inclusion of charts, graphs, illustrations or other visuals. Are you required to follow a specified format? Are there any complex ideas that could be clarified through the use of tables or charts, e.g., numbers, instructions, systems, or specifications?

2. **Are there additional design considerations that will influence the production/composition of the document?**

3. **When is this document due? Are there any other deadlines (drafts, etc.) that you need to consider?**

   List the exact days, dates and times.

Notes

Record any additional thoughts relevant to the project at the end.

Optional: Situation Analysis Memo

Your instructor may ask you to write a more formal **Situational Analysis Memo (SAM)** as a way to help you plan your documents and also as another instance of simulated professional writing. In other words, the Situation Analysis Worksheet is intended as an informal prewriting exercise. The SAM, on the other hand, is a formal representation to your instructor of the results of informal prewriting you did using the worksheet.

1. Using the **Memo Template**, translate the information on the Worksheet into a narrative memo.
   - You will notice that there are boldfaced headings on the SAM Worksheet. Use these headings in your narrative translation of the Situation Analysis Memo.
   - Plan on spending enough time on the Worksheet to compose an effective SAM. The more time and effort that you invest in completing the worksheet, the more you will benefit. You will make more economical use of your available time, and your writing will be concise and focused and therefore more effective.
2. Make decisions about the type and the amount of information that you include in the SAM. You will not need to include everything from the Worksheet in the SAM. Some of the information on the Worksheet will only be useful to you as you draft, revise, proofread, edit, and publish the project document(s).

Be sure to revise and edit the SAM. As a formal presentation of your work it is important to pay attention to both content and form.

Chapter 14

Project Assessment Memo

The Project Assessment Memo (PAM) should be a short memo (under 2 pages) that provides your instructor with an explanation of your approach to writing a response to a specific case. The goal is to persuade your instructor that your composing decisions were based on your critical analysis of the rhetorical situation of the case and were not just arbitrary or pulled out of a hat. You response should be based on your careful analysis of the case and understanding of the course principles introduced during the case.

> Use the PAM template, do not exceed two (2) pages, and include the following sections with these same section headings

Overview

Answer the reader's question: *What is this memo about?* Include a statement of purpose, a summary of main points, and a list of the items submitted along with this document.

Context

Answer the question: *What was your understanding of the project?* This section should include a summary of your understanding of the case including what your role was, what the situation was, and what your tasks were.

This section is also your chance to include information about how you approached the project that your instructor might not otherwise be able to get by looking at the documents you are submitting. It may also include details about the following:

- Assumptions you have made about the background or scenario
- Connections with the material or the project you may have that I don't know about
- Problems, conflicts, or contradictions you faced with gaps in scenario's background

- Information about the audiences your documents are addressed to
- An overall statement about your approach to a project or case

Documents

Answer the question: *What documents did you produce and why?* Subdivide this section by document, explaining the decisions behind each. This is the place where you do an analysis of the documents you are turning in and explain the decisions you made about them as well as how these documents relate to your overall communication strategy. Your decisions about purposes, argument, style, and format/layout are all possible topics. The idea is for you to explain why you have written the document as you have. Don't simply list the decisions you have made, you want to explain why you made them. For example, write: "The purpose of the Philips report is to X and Y. I wanted to X because….To accomplish X, I included details about…."

Production

Answer the question: What steps did you go through, *what strategies or techniques did you use to produce/make/write your final documents?* This section is where you give me details about how you produced the documents. Production generally leads the reader through the stages of development of the project. It includes such things as research conducted, planning or invention/analysis work done, accounts of group meetings, descriptions of group member roles and contributions, problems you had during the process, and ideas about how the production process may have influenced the work you are turning in. What new production strategies did you find particularly effective that you might use in the future?

Summary

Answer the question: *How did you accomplish/meet the project and course objectives?* This is the section where you review your analysis of the project. It is good to end with an evaluative statement regarding the success of the project. Remember to keep in mind the specific project objectives and the general criteria for professional business communication printed in your syllabus. Your assessment of the project should refer to these criteria and show how, specifically, your work meets or does not meet those criteria. Also remember to keep arguments/evaluations realistic; weak or overstated claims can actually do more harm than good.

Chapter 15

Definitions Project

> **Project Objectives**
> - Learn to write clearly by using definitions and other rhetorical strategies
> - Learn and apply principles of writing technical memos
> - Learn to integrate research into technical documents
> - Learn to apply ethical principles in the planning of technical documents
> - Gain additional practice using electronic writing technologies

In this project individual students research, plan, and write a policy memo for a fictional company. Beyond simply summarizing policies discovered through research, students must develop a *persuasive definition* that satisfies an organizational need for clear policy and that minimizes any resistance to such policy. In developing a usable policy for all parties you must carefully consider and apply ethical principles as well. In this way, you'll learn how clear and accurate definitions help readers understand and evaluate complex technical and social issues.

This project is a good exercise for business students who hope to be involved in the development and management of corporate policy. For students preparing for technical careers this project is a good exercise in clarifying complex information through definition, expansion, and example. Technical writers often convey complex, highly specialized information through the same strategies of definition (see Definitions chapter). Any business or technical document must be conveyed in language and concepts that readers understand, and learning how to write definitions helps writers translate complex ideas in reader-friendly ways.

Background: Bad Culture – New Management – New Policies

For this project assume that you work for a medium-sized software development company with 300 employees. The company has recently undergone several changes in upper management due to a company buyout and some rather nasty internal problems related to sexual harassment, corporate malfeasance, and media relations.

Several former employees threatened to sue the company amid allegations of sexual harassment. Women workers claimed that the previous management had fostered a hostile work environment. Among the allegations is that workstations were sabotaged, causing equipment malfunction and loss of productivity. A small few complained of physical assaults and verbal abuse. The complaints were settled quickly, primarily because the company was not in compliance with federal and state regulations regarding sexual harassment. Specifically, the company did not have a sexual harassment policy in place. Beginning several weeks ago, the new management has tried to bring the company into compliance by sponsoring several workshops on sexual harassment and announcing plans to implement a zero tolerance policy. However, some employees have claimed the anti-sexual harassment training was treated like a joke, with some male coworkers making inappropriate and harassing comments during the training. Others took the comments as tension relieving humor. Many employees—including both men and women—are fearful, distrustful, or resentful that the company is becoming "PC" (politically correct) and overreacting to a few isolated incidents within an otherwise good workplace environment. Questions about the meaning of zero tolerance have fueled all kinds of speculation and innuendo about permissible behavior within the company.

Many employees feel the company has been poisoned by the old management, which seemed to purposefully mislead the rank-and-file about the financial state of the company. Most employees believe the buyout—and subsequent restructuring and layoffs—could have been avoided if the management had cared more about the well-being of its employees than the profits gained by selling off a deliberately mismanaged company. This profit-at-all costs mentality has seemingly permeated the culture of the company. There exists more an attitude of competition rather than cooperation among employees. Distrust is the norm rather than the exception. The employees are worried about how the new management will treat them and about their futures with the company. In turn, the new management feels strongly that an ethics policy that covers issues of corporate malfeasance and employee conduct would help quell cynicism, repair morale, and foster a more positive workplace environment.

Another problem plaguing the company is bad public relations. The company earned its reputation in the IT industry for developing innovative spam and pop-up blocking programs that have been incorporated into many lesser-known e-mail and browser applications. It was only a matter of time before the company's products would be sought by the big players in the e-mail and browser markets. The company's future prospects roused the interest of many Wall Street analysts, business journalists, and IT industry executives. Employees found themselves inundated with phone, e-mail, and face-to-face requests for information about how its products work and what the future holds for the company. Some details of internal unrest related to sexual harassment and corporate greed also found their way into the mainstream media. The temptation to share insider information came to a head during the buyout, when some employees quoted in trade publications and national newspapers said too much. The company's new management wants a policy setting guidelines for employees dealing with the media and public.

At this stage the employees are confused, resentful, and paranoid about these issues. Your charge as a member of the communications group is to develop the company's new policies about sexual harassment, corporate ethics, or media relations (you choose one). Your specific task is to develop policy as a form of expanded definition. You have full backing from the new management, which wants to create a model workplace. Beyond compliance with legal requirements your organization seeks to improve relations between coworkers, management, and the public in hope of boosting productivity. Thus, you face an informative, persuasive, and ethical challenge: to move beyond the usual matters of clarity so that your definition promotes real understanding and reconciliation. You need to expand upon any legalistic definitions so that employees are able to understand clearly what does and what does not constitute inappropriate behavior related to any of the aforementioned issues. Thoughtless readers are likely to change their behavior only if they feel coerced. But appeals to fear almost always have limited success. You want your definition to convince people to accept the values that underlie the issues. This requires an expanded definition of each issue.

Online Resources: Writing Company Policy

"The Case for Written Policies" by Personal Policy Services, Inc.
http://www.ppspublishers.com/articles/written_policies.htm"

Guide to Writing Policy" by CSU Monterey Bay Policy Team
http://policy.csumb.edu/develop/write.html

Exercises

1. Research and find examples of company policies online. In small groups or as a whole class, discuss their common features. Which policies seem effective, and what elements contribute to their effectiveness? Which policies seem ineffective, and what elements contribute to their ineffectiveness? Develop a checklist of criteria for writing effective policy that covers content, organization, style, and format.

2. Write a sexual harassment policy memo that does not exceed 5 pages. Make sure the technical document is user friendly, both in content and form. Include a list of sources and/or copies of any memos you used as a guide or reference during your planning process.

3. If your class is doing exercise 2, have a discussion about how much can be borrowed from existing policies when you as a company writer are charged with writing a company policy. What might constitute plagiarism in this context? How should outside sources be cited in this context? What sources should be incorporated into policies and what sources shouldn't?

4. Conduct a situation analysis and/or write a situation analysis memo for this project. This can be done as a standalone exercise to reinforce situation analysis or in preparation for completing exercise 2.

Chapter 16

Cases: Overview of Goals and Strategies

Case projects ask you to assume a role in a fictitious situation (often based on actual events) and respond as if you were in that situation. Case assignments help engage the students in real world communication problems by simulating the ambiguity and political complexities of writing in workplace organizations.

Cases help develop the necessary critical thinking, problem analysis, and close reading strategies that are part of the writing process. Background readings will introduce you to principles of business writing, including ethical awareness and the conventions of formal business memos, letters, and reports. The cases give you the opportunity to apply these principles in simulated situations that are as close as a classroom setting can get to real world writing.

Students often find cases very challenging because they are being asked to think through ambiguous situations that do not have clear-cut right or wrong solutions. Background readings, class discussion, and instructor feedback should help you develop the critical thinking skills that lead to writing more effective solutions to such problems. The good news: if you mess up you can't be fired or sued as you might be in the real world. You can learn from your mistakes and be more prepared for future writing tasks.

Your instructor may also ask you to write a Project Assessment Memo, or PAM, that describes and justifies your response to the case. This is also a useful exercise, as it forces you to articulate your approach to the case. Good writers make strategic choices that are appropriate to a given situation. Writing a PAM helps make you a more deliberate and strategic writer.

> **Case Project Objectives**
>
> - Learn to see multiple purposes and audiences in particular contexts. A writer's choices have real-world consequences that go beyond the page. Cases are purposefully complicated and ambiguous: you need to draft documents that please your supervisor as well as maintain relations between your organization and the client company. These audiences may have different and conflicting values that, as a writer, you must discover and make decisions about. The act of writing begins at this stage of problem analysis and definition.

- Learn to see the organizational context. You're asked to consider who you are in each case. In other words, as a writer, you work in a complex web of relations involving overlapping and sometimes contradictory audiences. This organizational context puts the writer in multiple positions—employee, company representative, ethical agent to name a few. Your position changes as your audience changes: if you write to your manager, your position is quite different than it would be if you were writing to an external client.
- Learn to develop a communication strategy. Cases try to get you to see business writing as more than just writing a single text or document. Cases encourage you to begin to see how multiple documents—memos, letters, notes for interviews, research reports, presentations—come together in response to some need or exigency.
- Learn to establish credibility, trust, and goodwill through business writing. Maintaining goodwill is the most important strategy for all business communication. Coercion and hostility generally make business problems worse. You should certainly keep practical consideration in mind (i.e., money) as you think through case problems, but you should do so in a way that treats people fairly and avoids harm to others. Looking for win-win solutions is a cliché, but it will certainly help you navigate complex business problems where multiple interests are at stake.
- Learn to see generic features/format concerns of letters, memos and reports as rhetorical. Letters are primarily external communications, whereas memos are usually internal. Surface level features of the letter—headings, margins, salutation/closing—as well as clarity and tone all contribute to the persuasiveness of the document.
- Learn to use collaboration. Cases place you in collaborative situations, helping you see how seeking feedback from others is a valuable strategy when solving complex problems.
- Learn to use computer technology to design and revise documents. Many of the cases include MS Word templates that are consistent with the organization setting presented in the scenarios. You have access to these electronic files, which facilitate business writing in new but not necessarily easier ways. Remember to reflect on how the technology affects your project.

Questions for Analyzing Cases

These questions are designed to help you better understand the writing situations of each case. These questions can help you better understand the problem, anticipate possible solutions, and reveal any questions you or fellow students may have about the limited background information presented in each case.

Organizational Context

- Who are the **players** and what are their relationships? (Try mapping/diagramming)
- What are the lines of power within your organization and within the client organization?
- Where in the case do you see problems arising based on power?
- What information don't you know about the case that you have to assume? What information would help the situation? What are the sources of this information?

Rhetorical Context

- Who are you? What is your job? According to your job title, who do you think you typically write to?
- Why are you being asked to do this?
- What are your **purposes** here? How do you **prioritize** them? How can you accomplish them?
- Who should you write to (audience)? What are his/her/their interests and values?
- How many and what kinds of documents will you produce? What level of formality is suitable?

Ethical Context

- Do you agree with your boss? Is simply following orders the best possible response here?
- If you don't agree, what are your options?
- Who is likely to be harmed and how will they be harmed? Whose interests are being privileged? How okay are you with this situation?
- What is your socioeconomic background? How does gender, race, or class potentially factor into this case?

Focus on Managing Ethical Dilemmas

The complex, impersonal, hierarchical nature of contemporary organizations present some difficulties for communication *ethics*, the guidelines or values people should follow based on ideas of right and wrong. How does an individual mesh his or her personal ethic with the ethical norms and expectations of the organization and of society? Most ethical systems consist of the following two principles:

- **Respect others:** "Treat others as you wish others to treat you" is a maxim central to virtually every religious and secular ethical system. Some insist on a stronger version: *care for others*—work proactively to improve

others' well being, including the people you work for and with, the people you write to, and the people you write about.
- **Do good, do no harm**: Think about purposes of writing in terms of a larger good or an overall life goal. What good are you trying to accomplish? Why are you working? What are your goals—personal, social, organizational—and how does your work for an organization fit in with those life goals? What are the effects of your work?

These principles become very complicated in cases where unethical actions pose a threat or actually harm others. What is your position—what *should* your position be—regarding the following questions:

- Within organizations does responsibility and accountability reside at the top with the president, chairman, or corporate officer?
- Does responsibility reside with the immediate actors within the company?
- What ethical responsibility should be borne by managers and employees?
- Is the responsibility to be shared equally, or to varying degrees, by members of an organization?

When, for example, is it ethically responsible to "blow the whistle" on your own organization, on competitors, on clients, coworkers, etc? Is it acceptable to whistle-blow when you become aware of problematic actions or policies? How serious is the problem? Can it be resolved without legal or police action? Who should you "report" the problem to? How can you make superiors aware of your dilemma and not risk harm to yourself (your credibility, job security, etc.)?

The cases in this textbook will present ethical challenge to you, the writer. You will have to weigh whose interests should be served when considering possible solutions to the problems posed by each case. Keep the above ethical principles in mind as you deliberate on each case: (1) tread others with respect, and (2) do good, do no harm. Also, remember the first maxim of business communication: maintain positive relations, or maintain goodwill.

Online Resource – Problem Analysis and Planning

Use the following planning exercise to help you organize your thoughts during the problem analysis phase of each project:

Case Project Planning Sheet

Chapter 17

Big-1 Rental Agency Case

As the assistant director for human resources at the Pleasanton assembly plant of the Continental Car Corporation (CCC), you are responsible for helping the plant manager, Frank Page, in a variety of tasks. Today he calls you in to take care of a problem that has him visibly annoyed—probably more annoyed than you have ever seen him.*

"The Big-1 Car Rental Agency at the airport has really done it this time," Page says. "Yesterday we had a couple of VPs from a Japanese firm in here, along with an American who represents them out of San Francisco. After lunch, McConkey, the American fellow, and I were walking together out of the restaurant when suddenly he starts telling me about the terrible problems they'd had with the CCC rental car we'd reserved for them at the airport. Seems the heating system didn't work after the first five minutes, so they were freezing during the whole of the 45-minute ride from the Minneapolis-St. Paul airport. But that wasn't all. The car made some funny noises and just wasn't running well. Naturally, the polite Japanese didn't say a word, but McConkey didn't hesitate to say how unhappy and embarrassed he was about the heater not working. Well, if he was embarrassed, can you imagine how embarrassed I was? I didn't know what to say to the Japanese. What a fine way to advertise our products. Yes, sir. Let them see for themselves how wonderful CCC automobiles are!"

You shake your head. "They must have been miserable riding without a heater all the way back from the airport in the 10-degree weather…really impressed with CCC's quality!"

Frank Page nods. "And it isn't as if this is the first time that the Big-1 Agency at the Twin Cities airport has given our visitors lousy service. Several vendors have had problems lately, and have let people in the plant know about it. Why, Manuel Lopez was just complaining to me last week about the poor maintenance on the cars from the airport outlet. Some incident with the reps from the Zorelco Company who got a car that had ten things wrong with it. Lopez finally called in and had the agency come out with another car."

"Actually, the complaints have been coming in for several months at least," you reply. "I remember Joe Bomarito telling me some horror story about a high-mileage car from the Big-1 Agency about the time the addition was being put on the front reception area."

"That's a full seven months ago," Page says tersely. "Enough. We have a sufficient history of complaints on the airport franchise. Now we're going to do something about it."

"So what do you want me to do?" you ask. "Go see the manager at the airport agency?"

"No. There's no point in dealing with him. I believe he's proved his incompetence beyond a doubt. Any outfit that consistently rents poorly maintained cars is running a sloppy operation. I want something to happen. The best-made car in the world won't run well if it's not serviced properly. This Big-1 Agency is causing us an image problem."

"And of course we can't take our airport business anywhere else," you note.

"We have no choice," replies Page. "We can only rent CCC products at the Twin Cities airport by dealing with this outfit. That's why I want to blow the whistle on them. Make sure that they get their act cleaned up, NOW."

"So we go to the top, to corporate headquarters?"

"Right," Frank Page says firmly. "I want you to write a letter to the general manager of the Rental Division of Big-1 in New York City."

"You don't want to start giving him chapter and verse on our problems, do you?" you wonder.

"No, I sure don't. We haven't kept any records anyway. Look, Continental Car Corporation and the Big-1 Rental Company do a lot of business together. I just want to let the corporate people at Big-1 know that this is an intolerable situation and they'd better do something about it. Get their guys out here to look at this airport franchise and see how the place is being run."

"Got it," you reply as you head back to your desk to draft the letter for Page's signature.

Big-1 Case Notes

When formulating your response to the case, please keep in mind the following assumptions you must make:

1. **CCC Uses Big-1 Rental Exclusively.** Standard procedure at Continental Car Corporation is to rent CCC model cars for visitors from the Big-1 Rental Corporation. Big-1 is the major rental firm for CCC products across the United States. Like most automobile manufactures, CCC views renting its own brand of cars for visitors as a public relations effort, and so naturally expects the vehicles will be in top operating condition.

2. **What Frank Page is asking you to do is reasonable and within the bounds of your job** responsibilities. The first paragraph, in fact, states this: you "help" Page. Keep in mind that as an Assistant Director of HR you are part of the CCC managerial team at the local Pleasanton plant. You answer to Page, but at the same time, Page seems to think you capable of representing him and CCC to external business partners.

3. **You have all the names and contact information** of all the players in this case at your disposal. You do not have to consider gathering contact information as part of your response to this case. When writing your response documents, just add reasonable names and contact information to your documents. Use this information consistently and be sure whatever information you include conforms to the conventions of business writing (e.g., be sure to follow correct heading formats for letters and memos).

4. **You only know what is available to you in the case.** You cannot make up facts to suit you or to fill in the blanks with information absent from the case scenario. You cannot, for example, assume that you have received records from some other department within CCC or from Big-1 that proves someone's culpability in this case.

5. **Your response represents the next step in this case.** If there is some gap in the information, assume it is a legitimate gap that exists at the time of the case. Finding out information not available in the case might be one purpose of some part of your response. In line with Note #3, you cannot, for example, imagine you've received a reply to one of your documents, then formulate another document based on this invented reply. The documents in your response portfolio should represent a communication strategy, a set of documents that simultaneously attempt to address the situation.

Exercises

1. Write a formal complaint letter addressed to someone at Big-1 informing them of the problem. Frank Page asks you to write to the corporate headquarters. Decide if this is the best solution. Discuss other possible solutions and choose the most appropriate before writing to someone at Big-1.

2. Write a memo to Frank Page explaining a communication strategy that would best address the problem. This memo could include asking permission to gather more information and/or asking Page to reconsider his hasty plan, if necessary. You'll have to be tactful if you outline a communication strategy different than Page's.

3. Write a Project Assessment Memo (PAM) written to your instructor reflecting on decisions made in writing the case documents

Online Resource: Templates

Big-1 Planning SheetContinental Car Corporation Letterhead
Continental Car Corporation Memo

Big-1 Case Evaluation Checklist

Purpose: How effectively do the documents accomplish their intended task as defined by the writer and the multiple purposes of business writing?

- How likely will your response, the letter to Big-1 and memo to Page, address this problem while maintaining goodwill between
    - CCC and Big-1?
    - Page and CCC?
    - Page and Big-1 (local agency)?
    - Page and you?

Product: How well constructed are the documents?

- Do they follow the generic conventions of business letters and memos?
- Do they maintain a professional tone?
- Are they clearly and concisely written?
- Do they follow conventions of standard written English?

Production: How effectively was the document produced?
- Have all the required documents been produced?
- Have they been produced using the required templates?
- Do the documents reflect careful planning, drafting, revision, and editing? Does the writer's PAM reflect careful analysis and conscious effort to integrate course principles into problem analysis and document production?

***Source:** From *Cases for Technical and Professional Writing* by Barbara Couture and J. Rymer Goldstein (Boston: Little, Brown Co. 1985). Reprinted by permission of the author.

Chapter 18

A Business Faux Pas Case

> **Case Overview**
>
> The Business Faux Pas case is ultimately concerned with cross-cultural communication issues between a French and American company. But it also deals with ethical issues about what should be done in this situation, given the context and the tasks at hand. Obviously, tension between companies causes stress within companies. Bill Nestor, your immediate supervisor and family friend, has offended Madame Marie Jeaneaux, the President of an important French company (see Jeaneaux Letter). She sends a letter to your company's president and the problem finds its way to you, working in BellCom's Accounts Management department. To what extent is this case about cross-cultural communication, basic business etiquette, or poor judgment? You must decide.

You work as an account representative for BellCom Corporation, a growing telecommunications company that specializes in toll-free 800 lines, telephone debit cards, and data transmissions. BellCom is the nation's sixth largest long distance telephone company and among the fastest growing in its industry. Its annual revenue last year was 1,100,000,000, a 63.5% increase from the previous year. Its net income grew 47.3%, to $74,800,000. BellCom has been looking to partner with overseas companies that specialize in 3rd generation (3G) wireless technologies. You and your Senior Account Manager, Bill Nestor, have recently been sent to Marseilles, France to tour several facilities operated by Téléphone de Nice (NT), a major European 3G end-to-end systems provider (i.e., infrastructure, terminals, applications and expertise). This trip was the latest in a series of negotiations over BellCom acquiring the rights to market NT services in the U.S. as a BellCom product.

This trip was significant to you for many reasons. Not only was this your first business trip abroad, having been on the job only a few weeks, but it was also your first trip with Nestor, an experienced manager for BellCom who is also a good friend of your family. Your parents and Bill grew up in the same neighborhood together. Nestor's been close to your family for as long as you can remember. He was instrumental in securing your BellCom employment, under the condition that he would be your mentor. For you, it was a chance of a lifetime to secure this job and work under such an experienced businessman.

Nestor knows telecommunications and has had a lengthy tenure at BellCom. After working closely as a business associate of Nestor's, however, you've concluded that he's a brilliant buyer and manager, but seems a bit old fashioned in his values and ways of doing business. You've wondered on several occasions if Nestor was suited to overseas acquisitions. (Recently promoted to Senior Account Executive, Nestor's responsibilities have expanded from regional U.S. sales to all overseas accounts.)

During your trip to France you have actually become a bit disillusioned about Nestor as a mentor. It began at the French airport when Nestor said in a voice loud enough for others to hear, "Geez, do the French stink or what!" Even though several more such incidents left you quite embarrassed for Nestor, you decided that saying anything would be awkward.

After visiting the NT facilities as planned, you and Nestor were invited to join the brand new President of NT, Marie Jeaneaux, at a luncheon with several other French business representatives and managers. Expecting to arrive at a boardroom filled with platters of veggies, finger sandwiches, and cans of soda, all surrounding the imminent PowerPoint presentation, you and Nestor were shocked when your NT limousine took you to the world renown Palm d'Or restaurant. This would be a lunch unlike any other.

At the Palm d'Or, Marie Jeaneaux stood to greet her American clients and offered an accented, "Hello, how do you do," as she shook each of their hands. Nestor looked at you and winked then turned to Madame Jeaneaux with a wide, toothy smile and said, "What a lovely scarf you have on." A look of confusion fell on Madame Jeaneaux's face. An older man to her right said, "Bonjour Monsieur Nestor," and introduced himself, "Pontelle, Emmanuel," as they shook hands. "Nice to meet you, Pontelle," replied Nestor. After several awkward handshakes and "hellos" and "bon jours" around the table, lunch began. You recall striking up a conversation with a French woman sitting to your right and asking about the French Riviera. During this conversation you overheard Nestor ask Madame Jeaneaux, "So, what does your husband do?"

The waiter circled the table pouring red wine into each glass with care and grace. It took a few minutes to finish the pouring, and by the time he was done Nestor had finished his glass. No one else had yet taken a sip. You looked around and were sure there was tension amongst the French clients, despite the fact that their whispers and mumbles were beyond him. Not much of a drinker to begin with, you joined in Madame Jeaneaux's toast and took a sip.

Though the two-hour lunch was, as Nestor had told you, to talk about BellCom concerns, most of the time was spent eating and drinking. When Madame Jeaneaux asked you and Nestor if the food was enjoyable you replied, "Yes, it was delicious. Thank you." Nestor, with his mouth full, gave Madame Jeaneaux the OK sign with his fingers.

As the group stood from the table to say their goodbyes several French businessmen kissed Madame Jeaneaux on each cheek before departing. Nestor, following suit, faced Marie Jeaneaux and slurred, "Let me thank you for lunch by doing it your way." Then he put his hands on Jeaneaux's shoulders and leaned in to kiss her on the cheeks. Marie Jeaneaux pulled away in shock but didn't dare broach the subject with her inadequate English. Again, you watched the faces of Jeaneaux's employees wrench and gape in disgust. Not surprisingly, Nestor failed to pick up on the palpable tension in the room. When Nestor said, "Then we'll talk soon. Pleasure to meet you," Madame Jeaneaux simply nodded and walked out of the restaurant.

Though little business was done at lunch it seemed to have been a successful trip at the production plants. Nestor turned to you and said, "A job well done, kid!"

Two Weeks Later at BellCom's California headquarters...

Sherri Philips, Vice President of Operations, has asked to see you in her office. "Good morning," says Ms. Philips. "I assume you've seen the letter from Ms. Jeaneaux." (See Jeaneaux's Letter.) So you know she's threatening to back away from our company because of Mr. Nestor's behavior. I'm sure you know how important this deal is. If our company doesn't acquire rights to NT we'll be bought out for sure. We really thought that you and Nestor's tour would be a harmless way to get both your feet wet on this deal. And though we're certain you had less to do with this, you were there, which makes you involved in this mess."

You dip your head, showing some distress.

"You were with Nestor and your name is associated with the uproar. Tough luck. What we need from you is a brief report about what happened. I want you to figure out what Nestor did wrong and give me your recommendation on what the the company can do to avoid future problems in dealing with NT. Do some research if you have to."

"This document, along with Ms. Jeaneaux's letter, will be discussed in an executives meeting about Bill Nestor," Philips continues. "Look, Bernstein (BellCom CEO) and the other Vice Presidents are livid about this situation and want us to make amends and ensure this never happens again. I know you just started here, and we're glad to have you, really, but you need to help us clean this up or it could end up dragging you down, too."

"Nestor is to draft a formal letter of apology, and we'd like your name on it, too. I have already discussed this with Nestor and he should have a draft to you by the end of the day. I need both the letter and the report by Friday."

Later that day in your office...

"Can you believe that letter?" Nestor angrily murmurs. "I could be fired for what she said happened, you know that?" He pats you on the shoulder and hands you a draft of his apology letter(**see Nestor's draft**).. "Look kid, I screwed up but I need you to cover for me. This is the letter I want sent, so just put your name on it. And what's in the letter is what I'm going to tell Philips and Bernstein. So, please, make this what really happened, if you know what I mean." And at that point Nestor turns to leave. "Thanks kid," he says winking, "I'll see you at your folks' house on Sunday."

Jeaneaux Letter

12 Month 2003

Rick Bernstein, CEO
BellCom Corporation1
390 5th Avenue
Santa Monica, CA 90405

Dear Rick,

I am pleased to have had the opportunities to learn about your company and have generally been treated gently. As the new President, I must write to you to explain my concerns.

Weeks ago, you sent two associates of yours to tour our facilities. After their visit, we joined for lunch at the very nice Palm d'Or restaurant. The young one was very quiet and one woman said many nice things of him. The other man is the severe problem. Not solely was he rude in numerous areas of etiquette, like with the wine and food, but he made no effort to know my culture. When I come to the United States, Mr. Bernstein, I do not act badly like he did. I do not:

- Drink too much
- Speak badly of the food
- Invade personal space
- Ask personal questions that are sexist
- Talk about someone's dress

It is true that Americans and French have different ways to communicate, but I believe Mr. Nestor should have shown more respect. I am sure you should feel the same.

I very much respect you and your grand company, but I am not sure if Mr. Nestor is the right person to be representing your company during our negotiations. At the very least, I hope that he would receive some kind of education in conducting business in my country. I do not want to quit your company, but I will not be disrespected as a French person, as a competent businessperson, or as a female.

I strongly believe there are greater things in the future for our two companies. I am impressed with BellCom's growth potential, but I hope that you would take care to see how important understanding our cultural differences is if we are to move forward.

Sincerely,

Marie Jeaneaux
President, Téléphone de Nice

Téléphone de Nice
4 Rue de l'arbre, 6th Arrondissement
Nice, France 78911-3444tel: (1) 5564
5565; fax: (1) 5565 7848; Web site: www.NT.fr

Nester Letter

BellCom Corporation
1390 5th Avenue
Santa Monica, CA 90405
Tel (310) 458-8316
Fax (310) 458-8310
www.bellcom.com

Month 16, 2002

Ms. Marie Jeaneaux, President
Téléphone de Nice
4 Rue de l'arbre, 6th Arrondissement
Nice, France 78911-3444

Dear Mrs. Marie Jeaneaux–

In this letter I hope to make amends as well as address and respond to your accusations. First of all, I did not drink too much. The rest of the company members all shared several bottles of wine, during lunch I might add, and all I did was join in the festivities. That, I think is totally unfounded. Second, the food was fantastic and I gave you the "OK" sign when you inquired and I graciously thanked you for lunch, if you remember properly. When I asked about your husband, I was merely trying to get to know you. I apologize for wanting to establish a friendly business relationship. In America, we try to create a rapport. The only time I might have "invaded personal space" was when I kissed you goodbye and thanked you. I've seen enough of French culture to know that's how you all great each other hello and goodbye. You shook our hands, which was very "American" and so I tried to be "French" by doing the cheek kiss.

We hope your company will become a valued partner for many years, and I wouldn't want an overreaction such as this to cause a chasm between companies. [Your name] and I enjoyed our visit immensely and feel that we have been misunderstood; in fact, your misinterpretation of the events has jeopardized our positions at BellCom. We both agree that our behavior during our visit, and especially at the lunch, was appropriate and quite acceptable.

Please accept my apologies and know that we value your company. I won't be so friendly next time if you promise to be less paranoid. Obviously, you've earned your position at Téléphone de Nice and I respect that.

Sincerely,

Bill Nestor   [Your name]
Senior Account Manager   Account Representative

Exercises

1. Individually or in small groups, research and identify the *faux pas* that Nester committed during his trip to France. Based on your research, discuss strategies for avoiding cross-cultural miscommunication.

2. Write an internal report memo to Philips about the incident. You must write a report of what happened at the business luncheon where the offending incident occurred. How should you construct your account of what happened?

3. What are your responsibilities to Nestor, to BellCom, to Téléphone de Nice (NT)? Where, in the case, do you see problems arising based on authority, ethics, and communication issues?

4. Write apology letter to Mrs. Jeaneaux. You have also been asked to sign an apology letter drafted by Nestor (see Nestor letter). The letter is written from Nestor's point of view and may not be the most accurate or appropriate, but Nestor asks you to cover for him. You must decide what to do for both the apology letter and the report. Should the letter be sent as is?

5. Write a memo to Nestor that responds to his request to sign his letter draft. What if you believe the letter should be revised? How will you persuade him not to send his version and to accept yours?

6. Write a project assessment memo (PAM) addressed to your instructor reflecting on the decisions you made in completing any of the assigned documents.

Online Resources: Cross-Cultural Communication

- Business.com's cross-cultural communication directory listings
  http://www.business.com/directory/advertising_and_marketing/strategic_planning/global_marketing/cross-cultural_communication
- Executiveplanet.com http://www.executiveplanet.com
- Learnaboutcultures.com http://www.learnaboutcultures.com/
- Washington Library's list of books and other resources
  http://www.lib.washington.edu/business/guides/culture.html
- Worldbiz.com http://worldbiz.com/

> **Faux Pas Case Templates**
>
> Use the following templates as you draft your response documents
> - Nestor Letter/BellCom letterhead
> - BellCom memo template

Case contributed by Jenny Bania

Chapter 19

The Scanner Slip-Up Case

> The Scanner Slip-Up Case (see Case Information) puts you in a situation, based on actual events, where you have to repair a mistake made by the company you work for. Thomson, a computer products manufacturing company, has partnered with your company, Online Training Services, Inc. (OTS), to run a one-month Web-based training program aimed at the retail sales associates (RSAs) working in retail stores across the country. The program offers a free scanner to the first 500 individuals to score 100% on the training program's final quiz. However, due to a program slip-up, OTS sends the e-mail winning notification to 1,000 people instead of the first 500.
>
> Although the Scanner Slip-Up case arises from a software glitch, it is ultimately concerned with the precarious balancing act in the business world in which a company must decide how to correct its own error. This case also involves ethical decisions on the part of you, the communications manager, who as the voice of the company is heavily relied upon to smooth errors over through active communication with the clients.

You work as a Communications Manager for Online Training Services Inc. (OTS), a Web-based training company connecting computer products manufacturers to retail sales stores through comprehensive training solutions. The company's business rationale is that frontline retail sales associates significantly affect the sales of products, as most customers make their purchase decisions based on the feedback and advice given by point-of-sale professionals. Hence, it's good business practice on the part of computer products manufacturing companies to invest in the training of these sales personnel.

Bob O'Brian and Sheila Gallagher started OTS in Santa Fe, California in 1995 with a small but strong client list, two computers, and a rented studio that became their first office. Thanks to Bob and Sheila's initial connections within the industry, the company achieved a small turnover of $300,000 during its first year. With the reputation of the company spreading through the industry, its annual turnover in 2002 grew to $15 million. Its clients now include all the major names in both the computer products manufacturing industry—including IBM, Dell, Toshiba, Sony, HP, Intel, and Motorola—and the retail outlets—including CompUSA, Best Buy, Circuit City, and Office Depot.

The main service of OTS is a Web site platform created, designed, and hosted by OTS where companies offer training programs for the retail sales associates (RSAs). These programs are offered on a monthly basis and include comprehensive training courses, knowledge assessments (quizzes), and valuable prize incentives. The training focuses not only on product knowledge but the enhancement of selling skills.

OTS's company structure is headed by Bob and Sheila, whose titles are, respectively, Director of Marketing and Director of Business Development. It also includes a Communications Manager (your role), a Legal Manager, an IT Supervisor, sales trainers, and a group of Web content developers. As the company's Communications Manager, you report directly to Bob, and you handle all communications, including letters, promotional materials, and Web site content.

OTS is faced with a serious problem—one that could potentially make or break your growing position with the company. Thomson, the manufacturer of a wide range of scanners, has teamed up with OTS to offer an online training program for RSAs at top U.S. retail chains across the country. The participants in the program are required to complete a one-month training program and take a quiz upon completion. The first 500 RSAs to score 100% in the final quiz of the program will win one of Thomson's top-of-the-line flatbed scanners called the ELEGANCE 57.

OTS and Thomson have worked together in the past on smaller promotions and have established the beginnings of a strong relationship. This promotion marks an upward shift in the companies' relationship, in that Thomson is transferring a hefty sum of their marketing budget from their prior public relations agency and placing more trust in OTS. Thomson provides an undisclosed rebate to the retail outlets for each RSA that completes the training program. The online training program requires and average of 4 hours to complete. RSAs can complete the training in intervals or all at once; however, RSAs must earn a perfect score on the final knowledge assessment to complete the training. RSAs can retake the final knowledge assessment as many times as necessary. The average completion rate for such training is 98%. Thomson paid OTS $500,000 for the Elegance 57 scanner promotion, which enrolled 19,897 RSAs. The wholesale cost of an Elegance 57 scanner is $300.

Throughout the month of the contest you wrote promotional letters and materials. Now, as the contest is wrapping up, you discover a major error has been made by your IT manager. OTS has sent out 1,000 e-mail winning notifications to RSAs instead of 500, as initially planned. The e-mail notification (see E-mail Winning Notification) was supposed to be an automatic response sent to the first 500 RSAs that achieved a 100% score on the final quiz of the training program. This condition was intended to be built into the computer program. However, something in the program went awry and twice as many winner notifications went out. The error has been fixed from the technical end, and now must be dealt with from a P.R. angle, which is your job.

Upon discovery of the glitch, Bob and Sheila pull you into the conference room.

"What do you think we should do?" Sheila asks you.

You consider your options."Well, we could just keep quiet on the issue, purchase the 500 extra scanners ourselves and award them to the winners," you suggest.

"I'd rather not," Sheila says. "I'd prefer to use that as a final option. Perhaps we could write an e-mail to the 500 false winners, explaining the situation. Maybe we can offer them some kind of consolation prize instead."

"That might compromise our relationship with our retail partners," Bob interjects. "What if we wrote a letter to their marketing director admitting our mistake, but persuading her to double the prize total considering the success of the training program?" "Both of these options involve our admitting to the error," you point out to your directors. "It may affect our credibility in the future."

"Do you have a better idea?" Sheila says sharply.

"Why don't you think about it a while," Bob says to you. "Then I'd like you to draft a memo addressed to me analyzing our options and recommending what we should do. Also, prepare a draft of the document you think is a better option, whether it's a letter to Thomson or an e-mail to the 500 retail sales associates who won't be receiving a prize."

Bob and Sheila send you on your way, and you immediately consult with Leo, OTS Legal Manager. You explain the situation to him, and he gives the following reply: "Frankly, we're not under any legal obligation to grant the prizes," he says. "The e-mail notification was simply a public relations ploy, not a legally binding contract. It even says so in the small print. All we'd have to do to prove our case is print a winners list from the database showing exactly when each of the 500 winners completed the quiz." (See official rules.)

You walk away from Leo's office, your head spinning. You realize, despite your short time on the job, that this is the type of decision making that will help determine your success in the business world. The company has put their trust in you. You return to your cubicle, pull up a blank document on your computer, and ponder your options.

Official Rules: Thomson's Sales Training Program

The participation in the program is subject to the following rules:

Participants may register for training at OTS's Web site, www.ots.com.

The participant must be a U.S. citizen over 18 years of age and employed as a retail sales associate with an OTS recognized retailer (see the drop down list on the home page).

All assignments and quizzes in the program are mandatory for the award of the prize.

Any employee of OTS, or a family member thereof, is prohibited from participating in the program.

You must score 100% on the final quiz of the program to qualify for the prize.

Only the first 500 winners are eligible to win the prize.

The winners will automatically receive an e-mail notification. However, the actual award of the prize will only be made after documentation and verification on our end. The official documentation and a legally binding affidavit will arrive shortly after the initial e-mail. The decision of OTS remains final in the matter. Any disputes related to the prizes will be subject to the jurisdiction of the federal and state courts in Clark County, NV.

The delivery of the prize will be directly from the sponsor company and OTS is not responsible for any delay or damage in the delivery.

E-Mail Winning Notification

Date:_____

From: service@otsonlinetraining.com

To:_____

Subject: Your Thomson Training Prize

Dear (Participant's name),

Congratulations! We applaud you on completing the Thomson Sales Training Program, and we are pleased to inform you that you are one of the 500 sales associates all over the U.S. to win Thomson's ELEGANCE 57, the latest in Flatbed Scanners featuring dazzling clarity, speed, and durability.

By completing this exclusive training program you have further boosted your already strong selling instincts, and we are pleased to have you representing our products at the retail frontlines. We are happy to reward you for your dedication to the program. Please allow six to eight weeks for the delivery of your prize.

Congratulations again. We look forward to working with you in the future and wish you the very best in your selling career.

Best,

Communications Manager
Online Training Services Inc.

Exercises

1. Write a recommendation memo to Bob O'Brian, OTS Director of Marketing, that explains the best solution to the award notification error. Be sure your report explains OTS's options and recommends the one you think is best for the continued success of the company. The two main options—writing to Thomson or writing to the RSAs—involve an admission of error by OTS. Should you write a letter to Thomson requesting 500 additional scanners? Should you send a corrective e-mail to the false winners? Are there other options? What is the best solution to this problem? And what do you base your recommendation on? Remember, you cannot afford to harm your company's relationship with either Thomson or the RSAs, so you will want to choose your words very carefully.

2. Write a letter to Thomson that explains the error and informs them of OTS's solution. Before you write your letter, consider if you should you offer your audience any incentives to sell them your proposition, as is sometimes the case with adjustment letters. What will you write to establish credibility for the future, to ensure that both retailers and manufacturers continue to rely on OTS for their Web-based training needs?

3. Write a letter to the 500 retail sales associates who mistakenly received notification of winning the training promotion. Before you write your letter decide if it is best to withdraw the notification or offer some alternative solution.

4. Write a project assessment memo (PAM) addressed to your instructor explaining any of the documents you wrote.

Use the following templates as you draft your response documents
- OTS, Inc. letter template
- OTS, Inc. memo template

Use the following peer review guidelines to evaluate your drafts
- Peer review guidelines for Scanner Slip-Up case

Original Case Contributed by Kelle Schillaci and Anish Dave

Chapter 20

Insurance Fraud at MedTech Case

> This case puts you in the role of a human resources manager at MedTech, a company that manufactures medical supplies. It is reported to you that while an employee is earning compensation for a work-related injury, she is also running a bar, earning income, and performing duties inconsistent with her injury, including tapping kegs, cleaning tables and waiting on customers.
>
> Fellow employees within the company become angered when they learn that the company is investigating this woman's potentially fraudulent worker's compensation claim. Your task in this case is to justify the company's actions while building an atmosphere between employees and employer that is based on trust, mutual respect, and the achievement of common goals.

You work as a human resource manager for MedTech Manufacturing, a small but successful Midwestern medical equipment manufacturing company. MedTech produces 80% of the regional hospitals' waste equipment, including medical gloves, biohazard storage and disposal containers, catheter bags, IV bags, etc. You have just taken over for the previous HR manager, Bill Smith, who worked at the company for over 20 years. Since you are new to the area, you haven't met many people, nor do you socialize with the plant workers, most of who have grown up together, and are loyal, hardworking people. While MedTech is in a small town, it is a big business and a vital mainstay in the regional economy.

A female employee, Susan Seer, developed a work-related repeat trauma injury while working on a company assembly line manufacturing medical catheter bags. She strained the tendons in her wrist and sprained her elbow; since she assembles items on a line, this injury made it impossible to continue with her job until she healed. She went to the doctor, then applied for and received worker's compensation based on her alleged temporary disability, which amounted to a certain percentage of her normal income.

This employee also owns a bar and lives upstairs above the bar. She and her live-in significant other normally run the bar nights and weekends. It is reported to you anonymously that while she is earning compensation for being disabled, she is running the bar, earning income, and performing all normal bar duties, including tapping kegs, cleaning tables and waiting on customers: activities that indicate she is not, in fact, disabled pursuant to her physician's advice.

Under state law, the employee is supposed to report supplemental income, which would then reduce the amount of compensation she receives for the work-related disability. In addition, if it is found that the duties she performed in her tavern job are similar to her previous duties and ones that she legally claims not to be able to perform due to the injury, she may in fact be denied the work-related compensation altogether.

In order to establish evidence and get a first hand account of her activities in the tavern, you dispatch an insurance claims adjuster representing the company to visit the bar one evening to document her activities and review their potential as disqualifying factors in her compensation claim. He views the employee serving beer, cleaning tables, waiting on customers, etc. When the claims adjuster pulls out a video camera and begins filming the employee in the tavern, the adjuster is roundly beaten and thrown into the street by the customers in attendance. In addition, the adjuster's vehicle is vandalized and he has to take a taxicab home.

Despite this unpleasantness, the insurance claims investigator filed his paperwork and the insurance company stopped payment of Seer's injury compensation. Susan Seer, in turn, has written a letter to MedTech manufacturing, suggesting that she will file suit for if her benefits are not reinstated to compensate her for her work-related injury (see Susan Seer's Letter).

News of these events spreads rapidly on the factory floor, causing immediate employee relations problems. Fellow employees and friends of the individual in question are demanding to know why she is being harassed by the company, and what right the company has to send out an "agent" to investigate her. The employees are united in opinion and have little trust in you, the new Manager, to handle the situation. Among the comments muttered by the workers:

"Bill never would've harassed Susan like this!"

"Why is the company trying to cheat us out of worker's comp benefits?"

"It's this new guy causing all the trouble."

McKay, the Vice President of MedTech and your boss, has received Susan's letter and is understandably upset. He calls you to his office and says, "This Seer business is turning into a real mess. Why the hell is she collecting compensation and working? And if she's able to work, why hasn't she come back?"

"I want you to draft a memo to all our workers explaining why we had to take the actions we did. I don't want this to create unnecessary distrust among our workers. Tell them as little as possible but enough to make them understand why we have to follow the law."

"I also want you to draft a letter to Susan making it clear to her that she doesn't have a legal leg to stand on. We take care of our workers, but if follows through with this lawsuit she won't have compensation or a job to return to. I'd like you to write me a summary report once you've drafted the paperwork, and copy me on the memo and letter you send out."

You leave his office with a frown on your face. The sheer delicacy of the problem baffles you. This isn't how you wanted to establish yourself at the company, and your future may depend on whether or not you're able to gain the employees' trust.

Letter from Susan Seer

Susan Seer
821 Main St.
Waupun, WI 53963

January 15, 200X

Mike McKay
Vice President
MedTech
5000 Industrial Lane
Beaver Dam, WI 53916

Mr. McKay:

Your management has taken away my workers compensation benefits. Im not able to return to work for several more weeks as per my doctors orders. I am sure there has been some misunderstanding, but I need these benefits or I will not be able to pay my bills. If you are unable to do something to help me I will take you to court to enforce your payment of my benefits. It is my rights. I expect a timely resolution to this problem am look forward to hearing from you soon.

Sincerely,

Susan J. Seer
Susan Seer

Susan Seer's physician's report/certification of disability

PHYSICIAN'S AND CHIROPRACTOR'S PROGRESS REPORT CERTIFICATION OF DISABILITY

Claim Number: 22-0142
Social Security Number: 392-84-9574

Patient's Name: Susan Seer
Date of Injury: 8/20/04
Employer: Acme Manuf.
Name of MCO (if applicable):
Patient's Job Description/Occupation: Assembly Line worker
Previous Injuries/Diseases/Surgeries Contributing to the Condition: n/a
Diagnosis: torn ligaments, sprained elbow
Related to the Industrial Injury? Explain: yes — strain from repeat stress action
Objective Medical Findings: 8 weeks for muscle repair

- [ ] None - Discharged
- Stable: [x] Yes [ ] No
- Ratable: [ ] Yes [x] No
- [ ] Generally Improved
- [ ] Condition Worsened
- [ ] Condition Same
- May Have Suffered a Permanent Disability: [ ] Yes [x] No

Treatment Plan: 4wk Quinine prescrip, pill form, splint on left arm, no use of arm for 6-7 weeks related to lifting, carrying

- [ ] No Change in Therapy
- [x] PT/OT Prescribed
- [ ] Medication May be Used While Working
- [ ] Case Management
- [ ] PT/OT Discontinued
- [ ] Consultation
- [ ] Further Diagnostic Studies:
- [x] Prescription(s): Phenol 2x daily for 4 weeks

[x] Released to **FULL DUTY**/No Restrictions on (Date): 10/20/04
[x] Certified **TOTALLY TEMPORARILY DISABLED** (Indicate Dates) From: 8/20/04 To: 10/13/04
[ ] Released to **RESTRICTED**/Modified Duty on (Date): From: _____ To: _____

Restrictions Are: [ ] Permanent [ ] Temporary

- [ ] No Sitting
- [ ] No Standing
- [x] No Pulling
- [ ] Other: _____
- [x] No Bending at Waist
- [ ] No Stooping
- [x] No Lifting
- [x] No Carrying
- [ ] No Walking
- [ ] Lifting Restricted to (lbs.): _____
- [ ] No Pushing
- [ ] No Climbing
- [x] No Reaching Above Shoulders

Date of Next Visit: 10/13/04
Date of this Exam: 8/20/04
Physician/Chiropractor Name: Mike Knight
Physician/Chiropractor Signature: [signature]

D-39 (Rev. 7/99)

Online Resources

To help write responses to the issues in this case, you should research information about federal employee compensation laws, the legalities of company policy, and insurance company policies.

Start with sites such as the following, but be sure to do more background research if necessary:

- About Repetitive Strain Injury from BBC Healthy Living
  http://www.bbc.co.uk/health/healthy_living/health_at_work/physical_rsi.shtml

Resources for Worker's Compensation Laws and Issues:

- DisabilityClaims.com, a resource for long term disability claims and litigation, sponsored by the law firm of Riemer & Associates
  http://www.disabilityclaims.com/
- Wisconsin Department of Workforce Development's information on worker's compensation
  http://www.dwd.state.wi.us/wc/legal/default.htm
- Wisconsin's Law Library's information on Insurance
  http://wsll.state.wi.us/topic/insurance.html

News articles on insurance fraud cases

- Annual Report For Calendar Year 1999 Allegations of Worker's Compensation Fraud" by Wisconsin's Dept. of Workforce Development
  http://www.dwd.state.wi.us/wc/fraud/1999_Fraud_Report.htm
- "Roundy's Accoused of Insurance Fraud" by Tom Daykin, Milwaukee Journal Sentinel, 25 Apr 2004
  http://www.jsonline.com/bym/news/apr04/224825.asp

Exercises

1. Write a Situation Analysis Memo summarizing your analysis of the rhetorical situation for one or more of the documents you might write in exercises 2–4).

2. Write a letter to Susan Seer, the injured employee

3. Write a memo to all company workers tactfully addressing the issues

4. Write a memo to McKay, detailing how you approached the previous two documents. The documents require McKay's approval before they are sent.

5. Write a Project Assessment Memo informing your instructor of your approach to writing one or more of the documents from exercises 2–4.

Original case contributed by Heather Lusty

Chapter 21

Foodborne Illness on Festival Case

As the new Food and Beverage Manager on Festival Cruise Lines ship *Enamorada*, you organize and oversee dining accommodations for the 2,667 guests on each voyage. You report to the Cruise Director, Carrie Anne Causwell, along with other department managers.

During the current 9 day voyage to several Caribbean islands, more than four dozen passengers have suddenly come down with some sort of illness.

"We can't afford the negative publicity and potential lawsuits of another massive foodborne epidemic on this line," Causwell tells you. "We've got to act fast and contain this before it gets out of hand. Find out what in our food stores is tainted; Dr. Lovelett should have analyzed the samples by now. We need to pull everything that might contribute to other passengers contracting the same symptoms. Then, redo the remaining menus to work around the stocks you pull. When you've got the potential damage under control, write up a report[1] to our CDC representative[2] on how this happened and why it won't happen again, and let all the kitchen staff know what we're dealing with. And make sure they aren't talking to the passengers about it."

You nod in agreement, and begin thinking about the task at hand. First, you need to figure out what foodstuffs the illness sprouted from. Then you'll need to work around that for the remaining meals. You head to Dr. Lovelett's small office and lab.

"Hey doc, any idea what we're up against yet?" you ask. "Looks like the Norwalk Virus.[3] We must've picked up some bad shrimp. The passengers who've been sick so far have all eaten shrimp. It'll all have to be pulled; not even cooking it will minimize the risks to the passengers" he replies. Two meals a day on board feature shrimp and other shellfish prominently. You ask if this virus might also be in the scallops, calamari, clams, oysters, and other seafood. "Hard to say," Lovelett replies. "It's possible, but more than likely it's just the shrimp."

Guiltily, you remember the last minute scramble to fill food and beverage inventory. A combination of last minute bookings and underordering to cut food costs prompted you to okay a new vendor without checking them out thoroughly, although you knew it was a relatively new company and had no major contracts yet with other lines. This type of outbreak does happen, though, and it might not have been the product, but the kitchen prep and storage at fault. You've heard of Norwalk, but you're not entirely sure how it spreads.

Whether or not this is the kitchen's fault, how much seafood can you really pull before the menu gets bland and repetitive? On the other hand, if this outbreak affects more passengers, you could really get into trouble for not nipping it in the bud. A small incident would definitely do less public relations damage than an entire cruise full of sick passengers. Is it better to be safe than sorry? You've got to get to the bottom of the outbreak, do some quick, efficient decision making to contain it, and reassure over 2,000 passengers that they're perfectly safe, and then explain and justify your response to the CDC.

> **NOTES**
>
> 1. The CDC Vessel Sanitation guidelines require cruise ships to report illness any time more than 2% of its passengers or crew complain of gastrointestinal symptoms (e.g., vomiting, diarrhea).
> 2. The Center for Disease Control (CDC) runs a vessel sanitation program (VSP), which was instigated in the 1970s. This is a cooperative program between the CDC and the cruise ship industry to help minimize the risk of gastrointestinal diseases. It has had great success in minimizing outbreaks over the last 3 decades. Twice a year, VSP staff inspect over 140 participating cruise ships while they're in a U.S. port. Ships are scored on food, water, spas and pools, employee hygiene, and general ship cleanliness; only scores of 86 and above (out of 100) are considered passing.
> 3. Norwalk virus infection is an intestinal illness that often occurs in outbreaks. Norwalk and Norwalk-like viruses are increasingly being recognized as leading causes of food-borne disease in the United States. The viruses are passed in the stool of infected persons. People get infected by swallowing stool-contaminated food or water. Outbreaks in the United States are often linked to raw oysters. Infected people usually recover in 2 to 3 days without serious or long-term health effects.

Online Resources: Foodborne Illness

The following sources will help you consider the different audiences and language you'll need to use, as well as policies on how much should be disclosed to the government and passengers in cases of foodborne illness (FBI). Start with sites such as the following, but be sure to do your own background research as well:

CDC Vessel Sanitation Program (VSP) http://www.cdc.gov/nceh/vsp/

Dietary Management Association: Food Protection Article
http://www.dmaonline.org/fppublic/connect35.html

Water Quality and Health: Sea Sick Article
http://www.waterandhealth.org/newsletter/new/winter_2004/sea_sick.html

Cruise Ship Illness Lawyer and Attorney: Norovirus Blog
http://www.noroblog.com/promo/legal-services/

Exercises

1. Write a memo to the kitchen staff reporting on the incident. Your memo must inform the kitchen staff and direct them on how to respond.

2. Write a CDC incident report. Your memo must inform the CDC of the incident and your company's remedial actions

3. Write a letter to the passengers informing them of the situation.

4. Write a Situation Analysis Memo (SAM) addressed to your instructor in preparation for writing any of the documents you are asked to write for this case.

5. Write a Project Assessment Memo (PAM) addressed to your instructor explaining your approach to writing any of the documents you are asked to write for this case.

Original case contributed by Heather Lusty

Chapter 22
Job Search Project

Project Objectives

- Learn to tailor a resume and cover letter to a specific audience
- Apply page design principles to increase the readability and overall persuasiveness of the resume
- Write a persuasive cover letter that compliments the resume and enhances the overall application
- Integrate focused company research into the production of effective resumes and cover letters

In this project, each individual student completes the following tasks:

- Find an actual job advertisement for a position that you're reasonably qualified for
- Research the job and company thoroughly, using print and Internet sources
- Collect a minimum of three sources of information about the company, including one outside source (i.e., the company's Web site shouldn't be your only source of information about a company)
- Write a **Job Analysis Memo (JAM)** that articulates your researched understanding of the requirements of the position and how you best fit with those requirements
- Write a **resume** and **cover letter** following principles of effective job document writing. (NOTE: No Microsoft resume templates will be accepted for this project. If you want to base your resume on one of these templates, you need to justify the design choices in your PAM)
- Write a **Project Assessment Memo (PAM)** that describes and justifies to your instructor your completed job search project

This is not a project about how to write a generic resume. While the generic resume has a place (e.g., at job fairs or for online resume banks), sending out 100 copies of the same resume is one of the least effective ways of getting a job, because these resumes can't compete with a resume that was carefully prepared to address the qualifications of a particular employer. To tailor the resume and cover letter, you need to research the company beforehand, rank important qualifications, determine key words sought by the employer, and then present only the information about yourself that demonstrates you meet the desired qualifications and persuades the employer that you are among the most qualified applicants. The **Job Analysis Memo** exercise teaches the analytic process that should accompany the preparation of effective job documents.

Many students of business writing think they know how to write resumes. Most students already have a resume and may have even been required to write one for another class. But rarely are students taught the rhetoric of the resume, which is largely visual. The Resumes chapter that accompanies this project contains five basic page design principles in addition to covering the basic conventions of resumes. You'll learn to manipulate the page as a visual unit. Several kinds of visual design techniques, such as *the quadrant test* and *the vertical lines test* will be introduced, as well as page design vocabulary such as *white space, chunking, headings,* and *emphasis.* Visual design choices are rhetorical; they contribute to the persuasiveness of any document and are particularly important for resume and cover letter writing.

You may have never written a cover letter before, but it is customary to submit a cover letter along with the resume when applying to professional positions. The Cover Letter chapter that accompanies this project covers the basics of writing cover letters, including how the cover letter should compliment the resume by including specific examples that elaborate key qualifications merely listed in the resume. Similar to an interview situation where an employer asks for an example or two of the applicant's relevant skills, the cover letter should go into more detail with specific examples. Since the resume is written in outline form, it can't go into as much detail as the cover letter can.

This project emphasizes the process of research and analysis that accompanies the preparation of effective employment documents. In this project, you will tailor your employment search to a real organization, for a specific job. Most experts believe that the most successful job seekers are those who target a handful of specific companies at a time, adjusting their documents to suit the needs and requirements of each particular audience. More typical strategies, like answering classified ads or mass-mailing generic resumes, are less effective.

For the success rates of various job hunting strategies, see Mark Poppen's "Best and Worst Methods," by URL:
http://www.interbiznet.com/hunt/archives/990117.html
(scroll down to Jan. 11, 1999 article).

Find an Actual Job Advertisement

> Don't know what you want to be when you grow up? Here's three strategies for finding a job advertisement for this project:
>
> 1. Identify job titles for jobs you're interested in by searching on job-hunting Web sites and looking at career planning information.
> 2. Consider what kinds of work you'd want to devote your life to get paid for. Too many students rule out career in entertainment, music, and sports because the think the chances of breaking into those fields are too remote. Well, now's your chance to dream, and find that ideal internship or entry-level job.
> 3. Browse the U.S. Department of Labor's *Occupational Outlook Handbook*, which details various industries, giving job descriptions and prerequisites, working conditions and industry prospects, and includes information about salaries and promotion opportunities. URL: http://www.bls.gov/oco/

You must first identify an actual advertised position related to your professional goals. There are three basic paths to consider when choosing the type of project you want to pursue:

- **Entry level**: If you're a senior and graduating shortly, you can use this project to prepare for a post-college job related to your degree.
- **Internship/scholarship**: sophomores and juniors, you won't meet the minimum qualification for an entry-level job, a college degree. Since employers prefer to hire college grads with internship experience, sophomores and juniors should use this project to prepare applications for internships and scholarships.
- **Graduate school or promotion**: If you are planning on going on to graduate school or already have a job, you have a couple options in this project. You can use this project to prepare a graduate school application, writing a resume and personal statement. The Researching Graduate Schools box contains resources for researching graduate schools and writing personal statements. The graduate school option is for seniors only, since younger students don't have enough college experience to write a convincing personal statement. If you have a job already, you should look for a better job, either within your company or at another company.

Researching Graduate Schools

- US News & World Report's (usnews.com) "education" section includes graduate school rankings and other helpful information.
- PhD.org ranks program, publishes news articles, and other resources
- MBA.com is a good resource for those seeking a masters of business administration

Personal Statements for Graduate Schools

Personal Statements for graduate applications are similar to cover letters, but there is more emphasis put on the writer's qualifications. The format is also slightly different, as personal statements are written more in academic paper format and not usually in business letter format.

- "Writing the Personal Statement" by Purdue OWL: URL: http://owl.english.purdue.edu/owl/resource/642/01/
- "Personal Statements and Application Letters" by Indiana University. URL: http://www.indiana.edu/~wts/pamphlets.shtml
- "Personal Statement Worksheet" brainstorming guide by U of Central Florida Writing Center. URL: http://www.uwc.ucf.edu/Grad%20Gateway/getting_in_to_grad-school/worksheet.htm

You must locate an actual job advertisement, using either online sources, such as UNLV's Career Connections Web site, or more traditional sources, such as notices from your academic department or current employer (see Figure 22.1). Don't select a job where you clearly do not meet the minimum qualifications or would have to fabricate credentials. While you are not required to actually mail in your application, that obviously could be one useful outcome of this project. The more real the position is for you, the more effective your job documents will be.

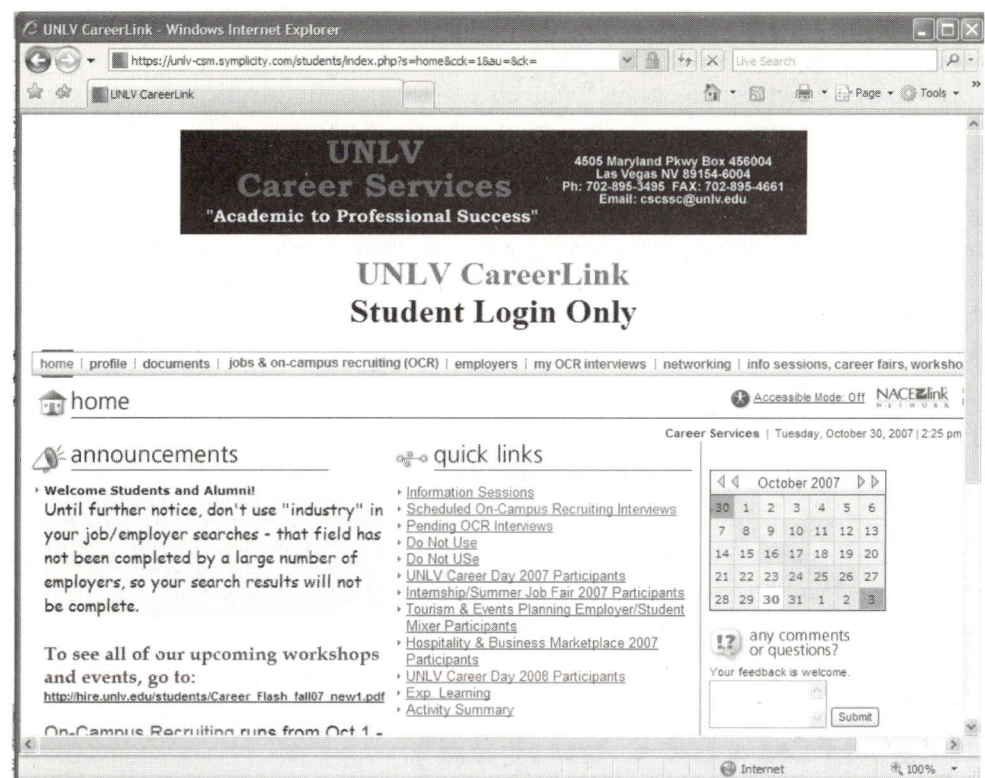

Figure 22.1. *UNLV's CareerLink. A job-hunting resource devoted to UNLV students, CareerLink includes a database of jobs and other tools, such as a resume posting and on-campus employer recruiting sign-up. URL: http://hire.unlv.edu/students/login.htm*

The Internet as a Job Seeking Tool

- **The Riley Guide.** Excellent resources on using the Internet as a job hunting tool. URL: http://www.rileyguide.com/index.html
- **Internet Job Hunting Dos and Don'ts.** Short summary of how to use the Internet in the job search. URL: http://www.quintcareers.com/Internet_job-hunting-dos-donts.htm
- **IBN's "First Steps in the Job Hunt"** Incorporates a daily tip with loads of archived articles on traditional and on-line job searching, from writing resumes, to interviewing, to identifying hot jobs. URL: http://www.interbiznet.com/hunt/

Chapter 22: *Job Search Project*

If there is a company you would really like to work for, yet it is not advertising any openings, you can still target the company, submitting your resume and cover letter as an unsolicited application. If you choose to write an unsolicited application, you still need to find a job ad for a similar position at another company, which will help you analyze the requirements for your unadvertised position.

If you are absolutely content with your job status, you still need to learn how to research and write effective job documents: Research suggests you'll change jobs between 5 to 15 times in your lifetime. Choose a realistic job that you *could* apply for, treating this project more as an academic exercise. Also, please *do not* choose a job you have held before, and *do not* write to an employer you've already met.

If you are unclear of the best path for you, talk about options with your classmates and instructor.

Your goal in this assignment is to design documents that will persuade an employer to grant you an interview for a professional position. But remember this: job documents usually never get you the job; they get you the interview that may or may not get you the second interview (and so on) that may eventually get you the job.

Research the Company

In today's tight job market, you need every advantage possible. One method for doing this is thoroughly researching the company or companies that you're most interested in. Anything you can learn about a company will assist you in designing a resume and cover letter that make you appear to be the most qualified person for the job. In fact, research shows that targeting 10 to 15 companies at a time through formal research and follow-up is a more effective strategy than mass-mailing your resume to 100 companies or posting your resume on an Internet job bank.

To develop your application for a specific company, you will need to research the company to find out about the job, the working environment, pay scale, benefits, and so on. Look for the following to help you write your resume and cover letter:

- Job ad with description of responsibilities
- Job pay, benefits, promotion opportunities
- Information about company from the company, such as a the following:
- description of company services or products
- mission statement
- history
- career opportunities
- annual reports
- press releases

Outside research

You should not rely on the company as your only source of information, since it is a biased source. Few companies will promote negative information about themselves. That is why you should also look to outside sources of information such as recent Internet, newspaper, or magazine articles about:

- What the company has done lately
- What's going on in the industry
- What the prospects are for the company
- Why this is/isn't a good company for you

Industry or career research

You won't always find lots of information about a given company, especially if the company is small or new. In cases like this, you'll have to be more creative, looking for information about:

- Similar or competitor companies
- Industry trends, or issues that all companies like your company are dealing with
- Career trends, or what general issues face people in similar positions as the one you're targeting

Once you've collected a minimum of 3 sources of information, including at least 1 source from outside the company, use them to **determine what the company wants**:

- What are the minimum qualifications needed for that specific job?
- What additional qualifications are desired for the position?
- What additional employee attributes are desired by the company?

A primary feature of the job search process is to critically analyze ads and other company information. Try to look for key words or concepts in the text as clues about what the company is looking for and values. In this way you attempt to identify specific adjectives or skills from the company literature to answer the following questions:

- What do the key words and concepts imply about the company?
- What do the key words and concepts imply about the company's goals?
- What do the key words and concepts imply about employee/company relationships? About employee/employee relationships?
- What do the key words and concepts imply about the company's expectations for employees?

Once you've developed a better knowledge of the desired qualifications for that particular position based on your analysis of the job ad and company information, **match your background to what the company wants**. Connect your skills and experience to the qualifications sought by the company. Ask yourself:

- What qualifications is the company looking for (what do they want?)
- How do I match the minimum qualifications for this position? (do I have it?)
- Do I have any additional, desired qualifications? (What are my strengths? What sets me apart?)

Completing the **Job Analysis Memo (JAM)** for this project will help you synthesize your research and articulate a deeper, more informed understanding of the rhetorical context of your targeted position. Does this mean you'll have to write JAMs for every job you apply for? No, but you should go through the same research and analysis process. Once advantage of writing your analysis down is that you'll have notes you can consult as you prepare your job documents and, hopefully, later for an interview.

Job Analysis Memo

The first document you will produce in this project is the Job Analysis Memo, or JAM. The JAM provides a formal opportunity to articulate the results for your job and company research. An effective way to synthesize, or make sense of, researched information is to put it into your own words. By thinking on paper, you will increase your chance of remembering it. You will also have a record to refer to later, such as when you prepare for your job interview!

After you've completed your company research and gathered a minimum of 3 sources of information (including one outside source), begin planning and writing your JAM. In memo format, include the following sections (there is no specific template for this assignment; use the intro memo template):

- Overview
- Job Description
- Company Description
- Important Qualifications
- Job Analysis
- Submission Format

Job Description

Answer the question: What will you do if you get this job? This section summarizes, in your own words, the responsibilities of the advertised position. Be sure to translate any technical language or professional jargon. For example, if a "Test Engineer" job ad states that "You will be responsible for application testing, and certification of new SAN (Storage Area Network) configurations, while working closely with both sales and development teams," don't just cut and paste that statement, but explain—to yourself and your instructor—just what's involved in "application testing," "certifying new SANs," and so on. If your job ad doesn't provide a detailed job description, you will have to consult additional sources, like other job advertisements or career planning information.

Company Description

Answer the question: What company would you be working for if you get this job? Do the same translating into your own words for the Company Description section. This information will come from both the job ad and the company research documents you've collected. Develop a detailed, concrete picture of the company: What is the company? What are its products or services? What is the company's mission? How big is it? Where is it located/headquartered? When was it established? How many employees does it have? How old is it? Who is its leader? What has the company done lately? What are the company's core values? Use this section to note anything about the company you've found that you think pertinent.

Important Qualifications

Answer the question: What qualifications are sought by the company? The final two sections of the JAM call for the hard analysis—call it soul searching—that will help you tailor yourself and your job documents to the "readers" at your targeted company. In the Important Qualifications section, summarize what you think the company wants, or is looking for. Prioritize what skills, abilities, knowledges, and values are most valued by that company for that particular position and in its employees in general. A well written job ad includes "minimum" and "additional" qualifications. What are they and which do you think are most essential to doing the job as described? What other qualities, not listed in the job ad, are valued by the company?

Job Analysis

Answer the question: What qualifications do you have? The final Job Analysis section is perhaps the most important. This is where you as concretely and as detailed as possible match "what you have," your skills and experience, to those desired by the company, "what the company wants." Be very specific. If the job ad asks for "3-5 years customer service experience," consider how you meet this criteria. Think carefully and critically about how what you've done or learned in the past might apply. Be creative and inclusive about fitting your skills to those sought in the job ad. Think about how you meet the minimum qualifications. What additional desirable qualifications do you have? Can you make a persuasive case that you're qualified for the position? What do you think you should emphasize in your job documents? Spend a good amount of time *explicitly connecting your skills to those listed in the job ad*. Answer the question readers at the company want answered: Why are you a good candidate for this position? The key is to base your answer on the criteria set forth in the job ad itself!

> *Writing tip: To prepare the final two sections of the JAM, make a two-column list labeled "**what the company wants**" and "**what I have.**" For each qualification you list under "what the company wants," brainstorm a list of as many skills and experiences you have related to that qualification as you can think of. Include concrete details about where you worked, for how long, specific projects you worked on, specific duties you performed, examples to illustrate soft skills like "teamwork" and "leadership" skills, etc.*

Submission Format

Answer the question: What format should you submit in your resume and cover letter? Does the position advertisement specify whether the application should be mailed, faxed, or e-mailed? Should the resume be formatted as a scannable resume or as a ASCII version? If e-mailed, should the resume be attached as a separate file or copied into the e-mail message itself? If you're given choices, which submission format will you choose, and why? Determine the actual submission requirements and adjust your plans for writing the cover letter and resume accordingly.

> **Checklist for the Job Analysis Memo**
>
> - Does the writer include the following sections: *overview, job description, company description, important qualifications,* and *job analysis?*
> - Does the writer identify and describe the job and the company (or division) in detail? Show knowledge of the company? Understand the responsibilities of the job completely? Understand the benefits fully?
> - Has the writer analyzed company promotional materials for skills and attributes highlighted by specific adjectives or key words in company literature?
> - Has the writer made explicit connections between the minimum qualifications of the job (from the job ad and company research) and his or her own qualifications?

Resume and Cover Letter

After you have determined how your background matches the qualifications in the position you seek, and thought about which of your qualifications you wish to highlight, you will then writer your resume and cover letter.

You should write your resume first. You may have an existing resume to work from, but you should quickly decide if it is in a format (chronological, functional, or combination) that suits this new position. You may need to start over with a new format. Also remember that a resume is an **outline of relevant qualifications**. That is, you can not simply update" an existing resume by adding new details to it, without removing irrelevant or old details, particularly if you are targeting a new field. If you have a resuem for you're your service jobs, you will likely need a whole new resume if you are applying for a professional internship or entry-level position. Non-traditional students starting a new career will also likely need to start a new resume from scratch.

As you plan your resume, follow the guidelines from the "Resumes" chapter in this textbook. You know you will have your contact information and objective at the top, but what details will you include in the body and how will you arrange those details. What resume format you choose will help you answer these questions. Do not worry about length or design at first. Err on the side of too much detail, but know you will have to cut down to get to 1 page eventually (for most resumes). Once you think you have added all the relevant details (and removed any irrelevant details), focus on visual design. Visual hierarchies can help you create easy-to-read levels of information. Use white space and chunking to balance the information vertically. Go for a simple, classic, readable "look." When you think you are done, seek feedback from peers, friends and family who have experience reading resumes, and set up an appointment for a resume critique at UNLV's Career Services.

Checklist for the Resume

- Is the objective statement specific about the job and company-oriented?
- Is the resume as a whole easily scanned? Targeted for a specific audience? Persuasive? Effectively designed? Using white space effectively? Pleasing to the eye?
- Do the sections support the claim(s) in the objective statement? Use lists instead of long sentences? Follow from most to least important? Order information hierarchically? Have a layout that helps the order of information?
- Does the writer use more than two fonts? Overuse highlighting? Give an address and phone number? Avoid overly dense sections? Select persuasive information? Adequately describe important experiences?
- Do the descriptive words support the claims for skills or knowledge? Use consistent structure? Focus on things important to the audience?

Since the cover letter highlights and elaborates key details that are listed in the resume, it should be written after you have drafted your resume. Remember that resume does not merely repeat what is in the resume. It elaborates the key qualifications with some specific examples that are either briefly listed on the resume or not listed on the resume. A good cover letter is like a substitute interview. Imagine the employer reading your letter and wondering how you meet his or her company's qualifications. The body of your cover letter is an answer to this question: "I believe I am qualified for X, Y, and Z reasons." But your cover letter does not just claim to meet the qualifications, it demonstrates or backs up the claims with concrete evidence, in the form of detailed examples, which are usually arranged in order of relevance. Your most convincing qualification should be list first (after the standard opening paragraph).

Just as with the resume, do not worry about length at first, but know you will eventually have to cut your letter to about three-quarters of a page. Saying all you want to say about yourself in less than one page is an excellent exercise in writing concisely. After you have the main details and organization down, then work on eliminating unnecessary words. Consider the cover letter's checklist and consult your peers and people who have experience reading cover letters. Career Services can give you feedback on your cover letter, as well.

> **Checklist for the Cover Letter**
> - Does the letter as a whole have all the necessary parts (return address, mailing address, date, signature and typed name)? Address a specific person? Attend to the audience's concerns and interests? Look professional?
> - Does the introduction state the specific job? Connect with the company? State the applicant's qualifications?
> - Does the body have focused paragraphs? Support any claims made about the writer or the job? Avoid merely listing experience? Persuasively describe qualifications? Make the qualifications clear?
> - Does the conclusion end courteously? Offer information for further communication?
> - Has the writer shown and used knowledge of the company? Carefully proofread the letter? Used appropriate language? Exaggerated any claims? Used her/his most relevant, persuasive experiences or qualifications? Presented him/herself professionally? Used an effective design/page layout?

Project Assessment Memo

Your instructor may ask you to write a Project Assessment Memo (PAM) for your Job Search Project. Your PAM should provide specific information on your approach to preparing your resume and cover letter. Include the following sections:

Overview: Answer the question: What is your purpose for writing the PAM?

Context of Project: Briefly review for your instructor what position you chose and why. You could mention whether or not you are serious about your choice.

Documents: What composing decisions did you make to write your cover letter and resume? You can subdivide this section into:

Resume: Discuss organization, content, design, and style choices. This section could consider questions such as: How does your objective statement focus on job and company needs? How did you make the document easy to scan? How did you target a specific audience? How is your design effective and persuasive? How do your sections support your claims? How did you hierarchies information? Why did you do it that way? Describe your font and highlighting strategies

Cover Letter: Discuss organization and content. This section could consider questions such as: What was the goal for your introductory paragraph? How did you connect with the company? What specific examples did you use to highlight your experience? Are these the most relevant and persuasive? Why? How did you focus the body paragraphs? How did you support your claims?

What persuasive tools did you use in describing your qualifications? What avenues did you open for future communication? What tone/style did you use? How did you show and use your knowledge of the company?

Production: How did you plan and write your documents? *Mention specifically what sources of information about your company you collected and what specific methods you used* (e.g., which library indexes and Internet search engines) to find your information. Discuss how this research influenced your planning and writing.

Summary: How well do you feel you met the project objectives? Evaluate how successful you were tailoring your job documents to a specific situation and perhaps what you learned from the project.

Links

UNLV's Career Services Center. Your new best friend, the local resource for area jobs and internships. Also includes information about campus recruiting visits and more. URL: http://hire.unlv.edu/

Career Service Center's "**Links**" page. An excellent compilation of links for researching job ads, careers, industries, and companies. Includes link to "**Career Link**," the local job database sponsored by Career Services Center. URL: http://hire.unlv.edu/links.html

Occupational Outlook Handbook U.S. Dept. of Labor site explores various industries, giving job descriptions and prerequisites, working conditions and industry prospects. Includes information about salaries and promotion opportunities. URL: http://www.bls.gov/oco/

WetFeet.Com includes everything from industry profiles and trends to detailed company information. Provides "insider" information from interviews with various company's employment directors. Also has career profiles, resume tips, and career advice.

Dick Bolle's JobHuntersBible.com. From the author of the *What Color is Your Parachute?* career-advice series comes this near-definitive clearinghouse of information about using the Internet for job hunting. Lots of information and links to all sorts of related help, all organized in a helpful format. URL: http://www.jobhuntersbible.com/

Career Journal from the Wall Street Journal. The job information area publishes new and timely feature articles daily. One of the best job sites out there. URL: http://www.careerjournal.com/

About.com's Career Page includes information about specific careers and general job hunting tips. URL: http://home.about.com/careers/index.htm

Yahoo.com's "Jobs" directory. Includes "most popular" and alphabetical listing of job search Web sites. URL: http://dir.yahoo.com/Business_and_Economy/Employment_and_Work/Careers_and_Jobs/Jobs/

Google's **directory of employment sites by industry**: URL: http://directory.google.com/Top/Business/Employment/By_Industry/

Google's **directory of resumes and portfolios by industry**. Look up sample resumes by field to see how others organize and present their skills. URL: http://directory.google.com/Top/Business/Employment/Resumes_and_Portfolios/By_Industry/

Chapter 23
Staff Development

Project Objectives
- Learn to use writing as a tool to prepare effective oral presentations
- Apply principles of visual design to presentation visual aids, including PowerPoint slides and printed handouts
- Learn to maintain professional development/communication skills

Staff development, otherwise known as training or in-servicing, is a common practice at most businesses. Many companies devote a significant amount of their staff development resources to building their employees' communication skills, from training in the latest industry-related computer software to raising consciousness of workplace gender and race issues. Translation: effective communication skills, broadly defined, are valued in the workplace.

For this project, you will be expected to work in a team to prepare and deliver a 10-12 minute presentation with accompanying visual aids (a PowerPoint slideshow and a handout). Your group's presentation will be a subject that interests you and that will benefit the class as a whole. You're encouraged to choose a subject that you're already versed in, from previous experience or casual interest, but you will also be expected to augment your existing understanding with some formal research from current published sources and/or expert opinions. However, being well versed in a topic area is not a requirement for joining a particular group. As a non-expert, you can offer valuable insight to the rest of your group as you prepare the workshop.

Your class will discuss possible topics. Presentations can (1) introduce some rhetorical principles/compositional practices related to business communication, (2) focus on instruction of a very specific feature of a particular piece of computer software, or (3) focus on describing communication practices in a specific field. Possible topics include:

- Instruction in computer software (e.g., intermediate to advanced features of Word; web authoring/HTML 4.0; making charts in Excel or Visio; preparing PowerPoint presentations; creating handouts in MS Publisher)
- Business communication trends (e.g., use of non-sexist language; use of portfolios in a job search; doing research on the Internet)

- International communication (e.g., cross-cultural communication)
- Influence of technology (e.g., rise of the e-mail genre or the future of broad-bandwidth comm.)
- Business communication in the small/medium/large company (e.g., exploring the differences)
- Remember that we're developing our "staff," so your presentation must include a practical element. In other words, your presentation must ultimately be deemed useful and interesting by your primary audience, your fellow classmates.

Even though business writing classes aren't speech classes, most of the basic rhetorical principles emphasized in this course apply to presentations in business contexts. In fact, you can't be successful without them. These skills include using writing and research to plan and organize your presentation; using effective design techniques for visual aids that support your presentation (e.g., handouts, overheads, PowerPoint slideshows) and using rhetorical principles for delivering presentations. These skills can be used, for example, to maintain your professional ethos at even the smallest, seemingly most informal of business meetings. Visual aids direct your audience's attention during such meetings and increase their retention of information afterwards.

Team Presentations Principles

The following chapters contain useful guidelines for planning and producing presentations and team-written documents:

- "Presentations"
- "Collaboration"

Deliverables

To complete the Staff Development project, each student will participate in groups to produce the following:

Component	Length and Medium	Elements
Staff Development Proposal	Length: 800 to 1,000 words, not including attachments. Submit print copy to instructor on due date	Team-written proposal must include: a **summary** of what you want to present an **audience analysis** a **rationale** for how the class will benefit a list of **outcomes** a plan for **visual aids** a list of background **sources**. Attach a detailed **script** including a presentation outline, plans for visual aids, designated speaker roles Attach drafts of PowerPoint **slideshow** and **handout**
Staff Development Presentation	12 minutes maximum (If you type a script, about two double-spaced pages equals a five-minute talk)	An effective presentation will reflect research, include clear intro/overview/body/ closing structure, be easy to follow (include clear transitions), and give practical strategies/tips that the audience (fellow classmates) finds useful and interesting
Visual Aids	PowerPoint slideshow and 1-page handout. You can use other visual aids as appropriate (e.g., projecting actual software application using overhead LCD projector)	Your visual aids will vary according to your presentation and will be more or less textual, based on role you see for your visual aid. Your visuals should, however, be appropriate for your purpose, professionally produced, easy to read, and leave the reader with key information, including sources
Project Assessment Memo	500 words or less individually written proposal	Evaluate each team member's contributions to the project, including your own
Audience evaluation	Completion of presentation evaluation for each presentation	Audience members are participants too. Failure to complete evaluations will hurt your final grade

Planning

> **Student Samples**
>
> Sample presentations available at companion website.

Your initial thinking about the rhetorical context for the Staff Development Project might start with the following:

- What will the audience, your classmates, find useful and interesting about your topic?
- What might your audience, your classmates, already know about your topic?
- How can they best understand your topic in a workshop environment?

Initial Writing Decisions

- Why are you developing these materials?
- What do you hope to accomplish?
- What is the purpose for this workshop?
- What are the benefits for using and understanding this material?
- What is the purpose of the handout?

Content Decisions

- What information must be present?
- What information would the audience find most useful?
- What major audience questions do you need to answer?

Design Decisions

- Which types of visuals would be most meaningful for your workshop participants?
- Which types of visuals would reinforce your expertise?
- How might you design the materials to maintain the participants' attention?
- How can you design a handout that the audience can use in the future?

Questions About the Rhetorical Context

- Are you an expert? How much do you know? What else do you need to know?

- How can you best show your expertise?
- How will you present yourself as both legitimate and valid?
- Who is the primary audience? What do you know about them?
- How much do they know about your topic?
- How can you reach them most effectively?
- What form should the materials take? Will they work together? Or will they be separate?
- How will you organize the materials?
- How can you assure that your materials will be understood?
- How can you assure that your handout will be useful?

Proposal

The staff development proposal is a collaboratively written document designed to help your team plan its presentation. The proposal also serves to inform your instructor of your plans and persuade him or her that your presentation has been carefully tailored to the audience (i.e., classmates) and will meet the requirements of the project.

In a 2- to 3-page memo addressed to your instructor, provide a detailed summary of your team's topic, a rationale for why it choose the topic and particular emphases of the presentation, an analysis of the audience's knowledge and attitude about the topic, a list of outcomes, a description of your visual aids, and a list of sources. Your team should discuss the design of its script, PowerPoint slides, and handout. You must also attach drafts of these documents to the proposal for the instructor's review.

This document is a proposal in the sense that your primary reader, your instructor, wants to know what you plan on presenting and how exactly you plan to do it. Thus, the proposal is not simply about a rough idea of what your team is thinking of presenting, but rather it should include a thorough and careful discussion of your presentation. This will enable the instructor to assess the effectiveness of your plans and provide feedback on anything that needs reconsideration or revision before your team presents.

Format

Your proposal should include the following sections:

- Topic Summary
- Audience Analysis
- Rationale
- Outcomes
- Visual Aids
- Sources

Topic Summary

Answer the question: *What will your team present and how will it present it?* Discuss the overall purpose of your presentation and the points you will cover to accomplish your goals. When planning the content of your presentation, remember that it should

- fit within the 10- to 12-minute time limit
- expand the audience's existing knowledge and skills related to the topic
- give practical advice that the audience finds useful and interesting
- be informed by outside research
- fit within the standard introduction/overview/body/closing structure
- be easy to follow, including clear transitions

The topic summary should review the basic outline of your presentation. It should also reference your script, including how your team divided speaking responsibilities.

Remember that a well written script provides the following information in an easy to read format: detailed outline, verbatim speaker dialogue, directorial cues, plans for visual aids, and a timeline.

Audience Analysis

Answer the question: *What is your audience's existing knowledge and attitude toward your topic?* Be as concrete and detailed when assessing your audience's familiarity and receptiveness to your topic. First develop a profile of the demographic makeup of your audience. Then, given this general profile, what can you assume they already know about your topic? What can you assume they should know or would like to know about your topic? Also, consider how receptive the audience will be toward your topic. If you think they may be hostile or disinterested, what might make them more interested?

Rationale

Answer the question: *Why will the class find this topic relevant and how did your team tailor the information to suit the audience?* For instance, justify why you choose to include certain information and exclude other information. Refer to your audience analysis when deciding what topics to cover in your presentation. Outside research may help you establish the importance of your topic for students.

Outcomes

Answer the question: *What should audience members know and/or know how to do as a result of participating in your staff development presentation?* This should be a list that articulates the objectives of your presentation. Such a list may or may not resemble the basic outline of your presentation. For instance, a presentation on how to use PowerPoint might include the following list of outcomes:

This presentation aims to introduce the class to the basics of PowerPoint, including the following:

- *how to use the interface*
- *how to create a slideshow from a template*
- *how to use animation effects*
- *how to avoid PowerPoint gaffes in business*

Visual Aids

Answer the question: *What visual aids does your team plan on using for its presentation?* Discuss in the choices your team made regarding visual aids, including what the particular design schemes are for the slideshow and handout, and why the team choose these designs. Discuss elements of design such as content, organization, format, style, and visual design, including use of graphics. Remember that the team is not limited to PowerPoint and a handout as visual aids. For instance, if the team is doing a presentation on how to use PowerPoint, the software application itself should be used as a visual aid. In this case, the team may elect to use PowerPoint visuals to introduce and overview the topic, but then ask students to launch PowerPoint at their workstation and follow along with a hands-on tutorial presented by the team.

Sources

The following are some online guides to Chicago Style:

- "Chicago Citation Style" by Long Island University, C.W. Post Library, URL: http://www.liu.edu/cwis/cwp/library/workshop/citchi.htm
- "Chicago Manual of Style Citation Guide" by Ohio State Libraries. URL: http://library.osu.edu/sites/guides/chicagogd.php

---

Answer the question: *What information did your team consult to inform its own understanding of the topic and decide what to present to the class?* You should provide **at least four (4) sources** your team referenced while planning the presentation. For the purposes of this assignment, use a *Chicago Style* format for documenting your sources accurately and consistently.

Proposal Evaluation Checklist

You can use the following checklist to evaluate the effectiveness of your proposal prior to submission.

Proposal

- Is the proposal in memo format, 2 to 3 pages in length?
- Does it include all required sections: overview, summary, rationale, audience analysis, outcomes, visual aids, sources?
- Does the summary describe topic and main points in sufficient detail?
- Does the rationale describe why this topic is useful to students in sufficient detail?
- Does the audience analysis describe who your audience is, what their current knowledge is about your topic, what their attitude is toward your topic, and what information you should emphasize based on your audience analysis?
- Does your outcomes section list in detail what skills and knowledge the audience will gain as a result of your presentation
- Do you have a list of 4 or more sources for your project (formatted consistently using a specific style for documenting sources)?

Script

- Do you have one master script that combines all parts?
- Does your script include an outline of main points, directorial cues for "who does what, when," verbatim dialogue for each speaker, timeline estimates, and text of visuals?
- Does your script plan for a 10-12 minute presentation? (No more or less)

PowerPoint Slides

- Do you have a title slide, overview slide, body slides, and a conclusion slide? (NOTE: software demonstrations omit body slides)
- Are your slides visually appealing and formatted following design principles of consistency, balance, visual hierarchy, and simplicity?
- Do your slides use text appropriately (size, parallel lists w/ fragments and minimal punctuation, readability, contrast w/ background)
- Do you have appropriate visuals/graphics where appropriate in your slides that enhance the text (charts, graphs, pictures, photos, etc)?

Handout

- Do you have a short handout that gives audience information they can take away from your presentation and reference later?
- Does your handout include a title, list of presenters, date, and context (e.g., business writing class)?

- Does your handout include details not appropriate for slides (e.g., big chunks of text, definitions of technical terms, complex charts or graphs, lists of sources for future reference)
- Is the handout visually appealing and formatted following the design principles of consistency, balance, visual hierarchy, and simplicity?

Purposes

- Does the proposal (incl. attachments) inform the instructor of your team's presentation plans and persuade the instructor that your group has a well thought out plan for its presentation?
- Does the proposal include evidence that the presenters will give a sufficient amount of information about the topic?
- Does the proposal include evidence that presenters are aware of the primary audience's (i.e., classmates') needs and interests?
- Does the proposal include evidence of thorough and appropriate research?

Product

- Does the proposal include the following sections: topic summary, audience analysis, rationale, outcomes, visual aids, and sources?
- Does the proposal reference within the document and attach to the document drafts of the following: script, PowerPoint slideshow, and handout?
- Does the script indicate a clear introduction, overview, body, and conclusion structure?
- Does the script contain an effective introduction that introduces topic, team members, and includes a hook to establish significance and relevance of topic for audience?
- Does the script contain an overview, following the introduction, that forecasts the main points of the presentation to the audience?
- Does the script use transitions effectively (e.g., "Now, Jane will discuss...)?
- Does the script include a summary before concluding?
- Is there evidence of plans to use adequate and appropriate visuals (slideshow and handout)? Are the visuals easy to read (appropriate fonts, use of specific headings, uncluttered design, balance of text/images/white space)?
- Does the handout include information, including resources/references, for take-away use?

Process

- Does the proposal, script, slideshow, and handout exhibit evidence of careful planning, drafting, revising, and editing?

Online Resources

Presentation Scripts

- *"Pre-Show Business" by Kevan J. Allbee, David L. Green and Kari Woolf.* Informative how-to discussion of presentation scripting: URL: http://web.archive.org/web/20040820074633 /http://www.presentations.com/presentations/search/search_display.jsp?vnu_content_id=1105023
- Sample Script: "Universal Access: Electronic Information in Libraries," University of Washington. URL: http://www.washington.edu/doit/UA/PRESENT/scintxt.html

Presentation Techniques

- *"Clockwork" by Dave Zienlinski.* Tips for managing presentation preparation time
http://web.archive.org/web/20060819200634/http://www.presentations.com/presentations/creation/article_display.jsp?vnu_content_id=1432490
- *"Developing a Presentation in Four Easy Steps"* by Presenters Online http://www.presentersonline.com/basics/delivery/4easysteps.shtml
- *"Presenting as a Team"*
URL: http://web.archive.org/web/20051214065212/http://www.presentations.com/presentations/search/search_display.jsp?vnu_content_id=1105192
- "Bad Delivery Habits." URL: http://www.presentersonline.com/basics/delivery/badhabits.shtml
- "PowerPoint: Boon or Bane?" By Abhay Padgaonkar. URL: http://www.presentations.com/msg/content_display/presentations/e3ife0aa2be553f374d10450eb2e6c43e02
- "Power Pointless" by Rebecca Ganzel. Discusses risks of over-reliance on PowerPoint's bells and whistles at expense of your message. URL: http://www.take-action.com/articles-powerpointless.htm
- PresentersOnline.com. This site offers advice from presentation consultants, plenty of tips and tricks, and a bank of great clip art to enhance the creation of slides. URL:http://www.presentersonline.com

Design Tips

- *"Quik Tips for Effective Visuals"*
URL: http://www.presentersonline.com/basics/visuals/effectivevisuals.shtml

- *"Tips for Text Heavy Slides."*
  URL: http://www.presentersonline.com/basics/visuals/ textheavyslides.shtml.
- *"Tips on Designing Brochures"* by Robin Williams
  URL: http://www.informit.com/articles/article.aspx?p=20718 &rl=1

MS PowerPoint and Publisher

- Microsoft's PowerPoint page
  http://www.microsoft.com/office/powerpoint/default.htm
- Microsoft's Using PowerPoint Links
  http://www.microsoft.com:80/office/powerpoint/using/default.asp
- Microsoft's Publisher page
  http://www.microsoft.com/office/publisher/default.htm
- Microsoft's Using Publisher page
  http://www.microsoft.com/office/publisher/using/default.htm

Chapter 24

Procedures Project

> **Project Objectives**
>
> - Learn to design, produce, and communicate clear instructions for doing a technical process
> - Learn to use writing as a tool for producing effective presentations, including writing scripts and designing visual aids (PowerPoint and handouts)
> - Learn new computer technologies that will be useful for this class as well as the workplace

In groups of 3 or 4 you will run a brief (~ 25 minutes) hands-on software tutorial workshop for the class. In addition, your group must provide written instructions for the steps you cover in class.

Most graduates in technology, engineering, technical graphics, and professional writing will be required to write instructions, or collaborate with others writing instructions. Writing effective, accurate documentation has both economic and legal ramifications:

1. The better the instructions, the fewer calls/e-mails to customer service staff.
2. Documentation that results in harm to users (either unclear or lacking proper warnings) is liable (every year, many companies are sued over faulty instructions).

The class will negotiate five or six software programs that most of the class is unfamiliar with but believes would be useful to learn. Your group will sign up to present a tutorial workshop for the one program that you are most interested in learning in-depth. You are not required to have mastered the software you will be presenting. This is your opportunity to learn more about software you're unfamiliar with.

Project Options

Your goal is to make the tutorials as interactive as possible. Rather than lecturing, have your classmates actually perform the tasks at their computers when possible.

Before choosing your topic for this project, make sure the software is supported by your school's computer lab. Since you need to plan a hands-on demonstration, you'll need to make sure that the application is available in your computer classroom. The following are some possible applications and functions that fellow students would find interesting and useful:

- **Intermediate to advanced Word functions**. Word has over 2,000 features. Some useful but little known functions include using tables, using textboxes, inserting graphics, setting page attributes, setting up form-fields (including mail merge), etc.
- **PowerPoint presentations**. Knowing how to choose and edit a basic slide show or how to use more advanced features such as inserting images and animation effects should be very useful.
- **Basic to advanced HTML code and/or HTML editors**. Your workshop could illustrate some basic codes, how to make simple pages, or how to use a popular editor. Make sure you choose a computer application that is available in your campus computer classroom, such as Dreamweaver, Netscape Composer, or even basic pages in Word or Notepad. This workshop works well when the entire class creates its own simple page and publishes it using their student computing user account. (Most school's have instructions for students hosting Web pages via their student computing accounts.)
- **Photoshop or Paintshop Pro**. Paintshop Pro is a less expensive alternative to the popular image editing software Photoshop. Workshop could demonstrate how to create and/or download buttons, backgrounds, and graphics for documents or web pages. Capturing and manipulating *screenshots* is a key function related to image editing software, one of great use in preparing the written instructions for the Procedures Project.
- **MS Excel**. Show students how to perform basic to advanced calculations or how to create charts and graphs and insert them into Word documents. This last feature will come in handy for the final Technical Report Project.
- **MS Publisher**. Show students how to create professional quality brochures, newsletter, reports, etc. using Publisher, Microsoft's desktop publishing software.

Deliverables

The following are documents your team will produce for this project:

- **Brief Proposal** in memo format outlining what program or function your team wants to demonstrate and why other teams should find it valuable. Your proposal also needs to assess who your audience is, including its needs, knowledge, and concerns about your topic. You will also attach drafts of your tutorial script, instructional handout, and your PowerPoint slides.

- **Tutorial Script** outlining exactly what you will be presenting to the class during your tutorial. The script should include not only who plans to say what and when, but also what will be shown on overhead/projector screen, checkpoints for evaluating whether or not your audience is keeping up, and other notes. The script should be attached to your proposal.
- **Instructional Brochure or Reference Card** documenting your tutorial. Your team will design a handout for reference during your tutorial and so the audience can remember and apply what your team teaches them afterwards. The handout must be carefully designed to fit a maximum amount of useful information into a relatively short amount of space/pages. You can decide between brochure or reference card layouts. Length of handout will be up to your team's discretion. Just remember, the longer the handout, the higher the risk of overwhelming your reader with information. But you need to provide enough information to have usable instructions, obviously. A draft of your handout should be attached to your proposal.
- **PowerPoint Slides**. While your primary visual aid during the presentation will likely be the actual software you are demonstrating and asking fellow students to follow along with, you will also want professional quality PowerPoint slides to help facilitate your tutorial, usually for the opening and closing of your tutorial. Drafts of any slides you plan on using should be attached to your proposal.
- **NOTE**: Each group must also submit final copies of their script, instruction handout, and PowerPoint slides the day of their presentation.

Steps for Completing the Project

Follow these steps to complete the Procedures Project:

- **Step 1: Do Background Reading**: Read the chapter on Presentations in this textbook. Since one of the main goals of this assignment is to learn how to compose effective instructions for print and oral delivery, you should also consult the online guides for writing instructions offered by Jonathan and Lisa Price, experts on technical writing and writing for the World Wide Web. These guides will familiarize you with basic concepts of writing instructions, such as determining how to divide a task into discrete steps, how to organize the steps, and how to write effective instructions.

> **Online Resources for Writing Instructions**
>
> - "How to Organize Step-By-Step Procedures" by Jonathan and Lisa Price
>   http://www.webwritingthatworks.com/CPATTERNprocedures.htm
>   - "The title is a Menu Item" by Jonathan and Lisa Pric
>   http://www.webwritingthatworks.com/DPatternPROC01.htm
>   - "Intros are Optional" by Jonathan and Lisa Price http://www.webwritingthatworks.com/DPatternPROC02.htm
> - "Put Instructions into Discrete Steps" by Jonathan and Lisa Price
>   http://www.webwritingthatworks.com/DPatternPROC 3.htm
>   - NOTE: The following are all linked from the bottom of "Put Instructions into Discrete Steps"
>   - "Format Steps So They Stand Out" http://www.webwritingthatworks.com/EPatternPROC01.htm
>   - "Write Short Energetic Steps" http://www.webwritingthatworks.com/EPatternPROC02.htm
>   - "How Many Steps?" http://www.webwritingthatworks.com/EPatternPROC 03.htm
>   - "Separate Explanations From Steps" http://www.webwritingthatworks.com/EPatternPROC04.htm

- **Step 2: Form teams and select a topic**: Two groups can present on a similar topic, but the presentations should not be identical, e.g., one group can present on PowerPoint basics and another team can present on intermediate or advanced topics. Which skills are basic and which are advanced depends on the group's estimation of the background knowledge shared by the majority of classmates.
- **Step 3: Do some background research**: Each team must cite a minimum of 4 sources in their proposal. If you're topic is PowerPoint, e.g., try using search engine terms such as *PowerPoint tutorial* or *PowerPoint basics* or *PowerPoint advanced* to find example instructions. You can use any samples you find as models to write your own tutorial, e.g., determining what are considered basic skills by the majority of tutorials you find.
- **Step 4: Analyze your rhetorical situation**: In a prewriting exercise and class discussion, consider the following questions:
  - Who is your audience?
  - What do they already know about your topic?
  - What should they know about your topic/what will they find interesting and new?
  - What topics should you include, given your audience and allotted time?
  - How should you arrange your topics, given your audience and allotted time?

- Who on the team will present which topics?
- What will the other team members do while one is presenting? (e.g., work keyboard, roam room and act as helper if people in audience gets stuck)
- How will the print documentation/instructions be integrated into the presentation (i.e., passed out at the beginning and used to follow along with, or passed out at end and used as take-away reference)?
- What should the print documentation/instructions contain?
- How should the visual layout of the print documentation/instructions be designed?
- How will the print documentation/instructions get produced?

- **Step 5: Draft your Proposal**: Your proposal for this project should inform your instructor of how you plan on presenting your tutorial. By the time you submit your proposal, your deliverables for this project—the proposal, script, slideshow, and handout—should be nearly, if not completely, finished and ready to go.

  Use the guidelines from the **Staff Development Project Proposal** to format and organize your proposal.

- **Step 6: Practice Your Presentation**: You can practice as a team nearly anywhere if someone has a laptop (provided it has the necessary software), or your group can meet in a campus computer lab and quietly rehearse your script.

- **Step 7: Make Any Necessary Adjustments**: Make sure your proposal fits within the allotted time limit. If when you practice as a time your presentation is too short, you will need to consider what other topics should be added. If your presentation is too long, you will have to decide what to cut. Revise your proposal accordingly.

- **Step 8: Submit Your Proposal and Practice Some More**: By submitting a proposal that represents as close to your final deliverables as possible, your instructor can review your plans and comment on whether or not any changes need to be made before the actual presentation day.

Chapter 25
Client Project

> **Project Objectives**
>
> - Write proposals, progress reports, and formal reports
> - Identify and solve organizational problems through writing and research
> - Plan and carry out a multi-stage, collaborative research and writing project (known as "document cycling" in the workplace)
> - Research the specific needs of the audience(s) to which you will be writing and tailor your writing process and products to these needs
> - Analyze and present a large body of information generated by your research
> - Design and draft effective documents and presentations, including final and intermediate oral and written reports, memos, letters, e-mail, and various types of visual representations of data
> - Manage a complex document production process which includes merging files, re-purposing existing documents, using style sheets and templates, and conducting multiple review and revisions

This project puts you in the role of a research analyst working to provide actual organizations—clients—with useful information. Student teams work with a client organization to identify a need or problem that can then be addressed through research. The final product of the project is a recommendation report that synthesizes relevant information (e.g., expert opinions, published research, industry trends) and offers feasible solutions. The project introduces you to the process necessary for producing group-written professional quality reports, including how to organize a team, manage a project, conduct primary and secondary research, and write preliminary documents that contribute to a final report.

What organization you work with and what problem you research depends in large part on the organizations and organizational needs identified by you and your client. Acceptable organizations include university organizations and local profit/nonprofit companies. Some student projects have included:

- Exploring ways to increase awareness, gain volunteers, and increase donations for a local animal shelter (Client contact: shelter manager, recruited by student)

- Recommending fundraising options to raise sufficient funds and to motivate member participation in UNLV's business fraternity, Alpha Kappa Psi (Client contact: club president, friend of student)
- Developing recommendations for a literacy program at an area elementary school (Client contact: student's mother)
- Comparing payroll processes and payroll software systems for a medium-sized electrical engineering firm (Client contact: student's mother)
- Developing a more structured and organized volunteer program for a nonprofit service organization (Client contact: student's boss)
- Evaluating web service options for a local apartment management firm (Client contact: student's boss)
- Researching web marketing strategies for an area mortgage company (Client contact: student's father)

For more help choosing a client, see the first part of the Planning section titled **Identifying a suitable client**.

The Client Project is a service learning activity. To learn how to adapt your writing skills to new contexts, the class will act like a consulting company. You will work together with your classmates and instructor to provide clients—actual Las Vegas organizations that have been selected by you—with information that will potentially help them provide better products and better services. The project creates an opportunity for real world experience, but at the same time provides the support and structure of the classroom. Unlike real world work situations, where the goal typically is to "get the job done," the main goal of this project is to help you learn to apply your writing skills to new situations. (Learning in the workplace is often only a secondary goal. Sink or swim, as they say.)

Your identity as a writer in this project is rather complex, but very authentic. You will primarily be representing yourself as a student doing a project for your business writing class. But since you will be interacting with actual members of the Las Vegas community, you also must be conscious of your role as representative of the University, the College of Business, the English Department's Professional Writing Program, your instructor, and your fellow classmates. You must also consider the values of your client organization, its clients or customers, as well as the greater Las Vegas community.

Deliverables

This project will extend over several weeks, focused on a research task performed for actual clients identified by you. It can include a variety of documents depending on the time allotted and instructor's discretion:

- **Client Proposal:** Each individual student is responsible for identifying an appropriate client and then writing a research proposal to the instructor. The class, with help from your instructor, will then choose the most appropriate projects for teams of 3 or 4 students to work on (i.e., from 24 or so proposals, about 6 are actually chosen as client projects).

- **Letter to Client**: After projects are selected, each individual student will write an acceptance or rejection letter informing his or her client about the status of the proposal.
- **Project Plan Report**: Each team will write a detailed plan for completing the research and producing the final report. The team may be asked to share the plan in an informal presentation to the rest of the class.
- **E-Mail Progress Report**: Each student will individually prepare, about mid-way through the project, a report that explains her/his individual and team progress within the project.
- **Final Client Project Report**: Each team will write a report that presents the client, the primary audience of the report, with the results of the its research and recommendations for addressing the client's problem.
- **Oral Presentation to the Client**: Each team may be asked to deliver a professional quality oral report based on the team's final written report.
- Project Assessment Memo: Each individual will write an individual e-mail PAM that evaluates how each team member contributed to the project.

> **Samples**
>
> Student sample documents for this project are available for review and classroom discussion at the companion Web site.

Planning

Planning the client project, like planning any informative report in business settings, is a complex process that involves:

- Identifying a suitable client (individual)
- Framing the problem or need as a research project (individual, group)
- Establishing a plan to manage the project (group)
- Determining how the final report will be produced (group)

Identifying a Suitable Client

Your first responsibility in the client project is to identify an actual organization that meets the criteria for a suitable client. Once you've found a client organization and obtained the contact's consent, you need to work with that client to define a suitable need or problem that calls for researched information.

Client Selection Guide:

1. List at least three organizations (businesses, nonprofits, clubs) that meets one or more of the following criteria:

- That you work for, belong to, or participate in (like a company, volunteer group, student club)
- That you used to work for or belong to
- That your family member or close friend belongs to
- That you would like to belong to or work for (If you're not involved outside of classes and your survival job, you need to get involved. Now might be a good time to make contact with a group that could help advance your career goals)

2. Which of these organizations do you have most **access** to?
   - Is there someone you are close-enough to that you can approach about this project and that, if selected, would be available for interviews? Local organizations are generally more accessible, but with e-mail and phones, it is possible to have high access with out-of-town clients.

3. Which of these organizations has a high degree of **willingness** to participate?
   - Who would be most interested and eager to volunteer their time to the project? Who would feel their organization could really use the help (as in, helpful information)?

4. Which organization will **benefit** most from the project?
   - Which organization has minimal resources and couldn't ordinarily afford to pay outside consultants for information? Nonprofits, small businesses, and student organizations are better choices than larger corporations.

5. Which organization can yield a project of most **interest and significance** to students?
   - Which organization would you and your fellow students find interesting/enjoyable/rewarding to work with or which organization could most contribute to you and your fellow students' educational goals (e.g., choosing a online record store because students are into music, choosing a finance company because many business majors plan on going into finance, or choosing any kind of company that has needs related to major areas, like a Youth Sports Little League that needs a fundraising strategy)?

In your proposal, you need to persuasively establish that your client is a good choice using the criteria identified above.

Framing a Problem or Need as a Research Project

Once you've identified a client and received informal consent from the contact, you should begin a fact-finding dialogue with them about how this project can help improve the mission of that organization.

1. List the services, products, mission, activities, or characteristics of this organization
2. Interview the client to determine which aspects of the organization could be improved

- Which aspects have problems or do not meet the expectations of the client, the client's customers, industry expectations, etc?
- Which aspects of the organization could be improved, rethought, or expanded? Research can be pro-active too, or help an organization to strategically improve its services, define new goals, or adapt to industry and technological trends. There's no such thing as a perfect organization.
- If you have trouble identifying a problem about the status quo, or the current state of the organization, begin by exploring what should be, or how to improve seemingly problem-free aspects of the organization.
    - *How* do other organizations do "X"? (e.g., use the web)
    - *Should* the organization do "X"? (e.g., have an e-business strategy)
    - *What* would improve "X"? (e.g., our current e-business strategy)

3. Define the Problem as a research question: Once you've identified a problem- or need-based topic, you need to articulate a researchable problem. A researchable problem is one that is framed as an open-ended **guiding research question**. The *how*, *should*, and *what* questions above are examples. Defining the problem as a research question helps you focus your research activity on seeking information that will best answer the question. This straightforward, systematic way of conducting research has helped scientists for centuries. Scientists use the scientific method to articulate a hypothesis in the form of a question that is then verified through experimentation.

Guide to Formulating a Primary Research Question

Write out three versions of a question focused on exploring the problem or need you identified. Try using different journalistic question words—who, what, where, when, why, and how.

How	Prompts procedural questions	"How can Acme improve its mail sorting system?"
What	Prompts comparisons and lists	"What electronic mail sorting system is best for Acme?"
Why	Prompts cause/effects questions	"Why is the Acme mail system not working?"
When	Prompts timing questions	"When should Acme upgrade its mail sorting system?"
Where	Prompts questions of location	"Where should the mail sorting duties be located in Acme?"
Who	Prompts lists of players, participants	"Who is affected by Acme's inefficient mail system?"

> - Be sure the from of the question includes reference to the client organization
> - Be sure the question is the *primary-question* (secondary questions are those questions that need to be answered before the primary, or overarching question can be answered)
> - Be sure each question is open-ended, or not easily answered by yes or no
> - Be sure the question is not biased, or implying a preferred answer (e.g., Why should Acme Insurance choose PowerPoint over Corel Presentations 10? *This question is biased; it assumes PowerPoint is the better choice!*)

4. Choose the form that you feel will yield the best information Once you've chosen a question, re-evaluate it using the criteria of access, willingness, benefit, interest, and significance. Be sure the problem is **manageable** in the time allotted for the project, that it is: (1) not too simple or too complex for a group of students, and (2) can be researched adequately in timeframe of the project.

You will need to argue the feasibility of the problem as you've defined it in your proposal using the above criteria.

Establishing a Plan to Manage the Project

Once you have been assigned to a client project team, which may or may not be the project you proposed, your group needs to work together to develop a plan to finish the project.

> Read the chapter on Collaboration, particularly the part titled Document Production for principles on managing the production of a team-written document

Your team will need to work together to write the **Project Plan** report. This involves developing:

- a research design plan
- a project plan.

Research Design Plan

> **Why research?**
>
> - To find out about the client's needs/problems (client research)
> - To find out client's ideas for solving problems or meeting needs (what will be acceptable to client) (client research)
> - To find out what is feasible/realistic for client (e.g. what can the client afford) (client research)
> - To find out about options that may help the client (external research)
> - To discover industry-related practices which address similar needs/problems (external research)
> - To discover industry-related trends which might influence needs/problem (external research)

Research is essential for addressing your client's problem or need. Research may have acquired a negative association for you in school from lots of research term papers that didn't seem worthwhile or enjoyable to you. But the fact of the matter is that anytime you need information, you are conducting research. Have you ever done any comparison shopping before buying a cell phone plan, car, or electronic device? You were researching. Businesses conduct research all the time to collect information about their target market, competitors, new products and services, human resources, etc. Businesses rely on information so much that people make careers as business analysts who specialize in researching data and making sense of it.

> **Developing a Research Plan**
>
> 1. Download the research design template to plan your research
> 2. Identify secondary questions that need to be answered
> 3. Develop a strategy for answering the questions:
>   - Primary sources (interviews, observation, surveys)
>   - Secondary sources (printed/published material)
> 4. Estimate how long each research task will take to complete
> 5. Assign research tasks to individuals
> 6. Create a Gantt chart depicting tasks and responsibilities

A research design plan is your team's plans for collecting information that will help address the need or problem identified by the client. At this state of the project, you team will brainstorm what information it needs to collect to answer the main question. This information can also be articulated as questions. For example, if your team is researching the question, "What is the best accounting software for a small company?" one obvious question that needs to be answered is: "What are the available accounting software programs for small businesses? You might need to answer the question: "What is my client company's criteria for choosing the software?" You might also want to know what other similar companies use and if there are other alternatives, such as hiring an accounting company. These are all **secondary questions** that need to be addressed to answer the main research question.

You also need to think about what are the best sources of information for answering your secondary questions. Should you look interview the client, look up information in trade journals or newspapers, interview experts, investigate competitors? As Table 1 shows, there are two main types of information: (1) information that comes internally from the client and (2) information that comes externally of the client, from first-hand interviews with experts or customers, to second-hand information published in magazines, newspapers, and journals.

Table 25.1.

What sources of information will yield best results?	
Client /Internal	**External**
Interview. Efficient way to get individual viewpoints. But reliance on isolated viewpoints can be limiting. **Survey.** Allows researcher to collect large number of viewpoints BUT time consuming to develop, distribute, and collate survey. **Observation.** Allows researcher to gain first-hand knowledge and provides means of verifying personal views ("what really happens" vs. what people say or think is happening") BUT can be limiting and time consuming. **Records.** Allows researcher to gain historical perspective on problem and provides "hard" evidence for certain kinds of problems BUT records don't always provide evidence of problems.	**Other companies (interview, direct observation).** To get information about what other companies are doing, to help the client. **Faculty/Professionals (interview).** To get expert opinions and suggestions for sources; strategies for doing research. **Product Vendors (interview, brochures, product literature).** To find out about available options and to get product information. **Internet (online research/browsing).** To track down on-line information of relevance to your subject; corporate websites may provide some information (or the name of a contact person for an interview).
NOTE: Never rely on only one source of information. Use at least a couple of different sources and a couple of different methods to cross-check and verify your understanding. Do as comprehensive a job of research as time and resources allow.	

After you team brainstorms its secondary research questions and determines the best sources of information for each question, it will next have to estimate how long it will take to answer each question. It may take several weeks to conduct a survey of customers, or it may take a few hours to interview the client. Don't forget the time it takes to prepare surveys or interview questions when estimating the duration of a research activity.

Finally, your team will need to decide who on your team will complete what research. The project planning phase will help your team when it comes to determine who will be responsible for particular parts of your group project.

Project Plan

A project is any multi-step process that involves the coordination of people, organizations, resources, and time. Project management and project planning grew out of the defense industry in the 1950s and 1960s as a method for planning and implementing complex projects that involve scores of people and often multiple companies.

> **Six Steps to Effective Project Management**
>
> 1. Establish the project
> 2. Build the project team
> 3. Organize the work
> 4. Assign tasks (and estimate resources)
> 5. Set the project schedule
> 6. Complete the plan

The common method for visualizing a project plan is called a Gantt chart, a horizontal bar graph that depicts a project's discrete tasks and the estimated time it will to take to complete each task. Major deadlines, called *milestones*, are depicted with symbols such as diamond or triangle. Because of the importance and prevalence of project management for getting information-based work done, there are lots of software applications available for creating Gantt charts and other project visuals, e.g., Microsoft Project.

> Gantt charts visualize project plans, including tasks, responsibilities, and time (milestones/ deadlines)
>
> - "Brief tutorial on Gantt charts"
>   http://www.me.umn.edu/courses/me4054/assignments/ gantt.html
> - "Creating Gantt charts in Excel"
>   http://pw.english.purdue.edu/resources/doc/gantt/index.shtm

Once your group has determined its specific research tasks, it should then identify the discrete deliverables that are due to complete this project as a classroom assignment. If your instructor is following this project as written, each team will have to produce the following documents:

- Project plan
- Client letter

- Progress report
- Draft of recommendation report
- Draft of presentation
- Final report
- Final presentation

When the research tasks are combined with the project deliverables, the project plan is complete (see Figure 25.1 below).

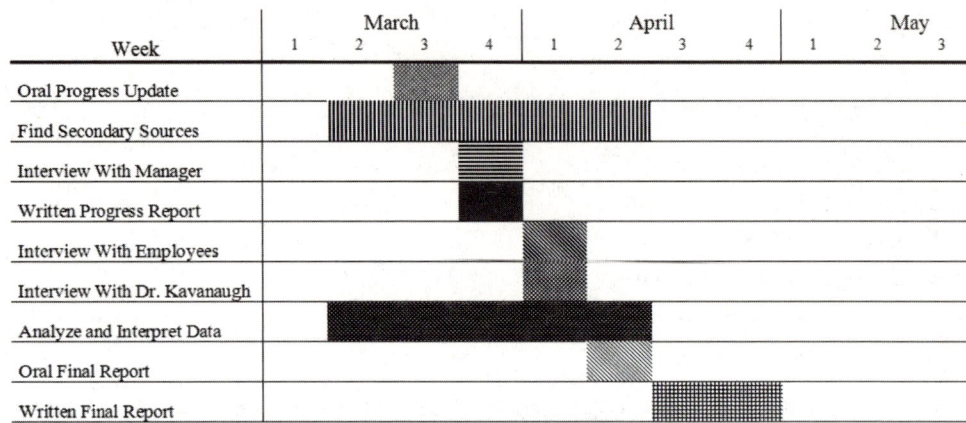

Figure 25.1. *Client Project Plan Depicted in a Basic Gantt Chart*

The **Project Plan** report will be a memo written to your instructor summarizing him or her of your group's plan for completing your Client Project.

Determining How the Final Report Will Be Written

How will your team coordinate the production of the final report? One of the biggest challenges is bringing the work of multiple authors into one, common format. To do this, your team needs to decide ahead of time issues about production responsibilities and deadlines, content plans, and style guidelines (e.g., how to format text, consistent spellings, etc).

Tips for Collaboratively Producing Reports

- Discuss general purpose and form of report
- Make an outline and develop headings and subheadings, include plans for visual aids (called a "storyboard," see below)
- Design a template/develop a "style sheet"
- Import "boilerplate" text from preliminary reports (proposal, plan, progress reports)

- Assign sections to various individuals (after group has collaboratively determined outline, including recommendations
- Write distinct sections individually
- Exchange sections and do revisions
- Merge separate files into one master file (and make backups!)

Storyboarding, like scripting presentations, is a practice borrowed from artistic designers. Like comic book, cartoon, television, or movie storyboards, report storyboards lay out in advance a detailed picture of a plan for the report, including the report's title, organization, section headings and subheadings, content, and graphics. It also indicates individual responsibilities.

Use the storyboard template to plan your final report

Follow the guidelines for writing business reports discussed in the "Reports" chapter of this textbook. Business reports are closer to published books than academic papers. That is, business reports often have front matter" elements such as a cover and table of contents. Business reports also pay more attention to visual design, as it is understood the visual "look" of the report contributes to the organization's image and the persuasiveness of the report.

You report must include at least two visuals in your report, including one table and one figure. This is an arbitrary number to push you to practice using visuals as a way of presenting information clearly and quickly. You can include more than the minimum.

Visuals are an important component to reports becaused they:
- Promote understanding/clarify complex ideas
- Add emphasis to key points
- Attract the eye
- Save space
- Arouse interest
- What deserves a visual?
- points that need emphasis
- points you want to draw reader's attention to
- points that need clarification (e.g., numbers, stats, processes, abstract ideas)

Chapter 25: *Client Project*

Remember that each type of visual is good for showing certain types of information, so think carefully about choosing the most appropriate type of visual for the information you wish to display.

> **What visual should you use?**
> - **Bar charts** show comparison and contrast
> - **Line graphs** show trends
> - **Pie charts** show percentages/parts of a whole
> - **Photographs/drawings/screenshots** give exact or facsimile representation
> - **Organizational charts** show structure or hierarchy
> - **Flow charts** show steps in a process
> - **Diagrams** show concepts and processes
> - **Tables** array information in rows and columns; good for showing lots of data, abstract juxtapositions, comparison/contrast (e.g., comparing several computers according to various specifications like cost, processing speed, memory, RAM, etc.)

Also remember that *visuals don't speak for themselves!* They must be explained in the text. Do not just write, "Here is a chart showing our third quarter profits." Instead write, "As Figure 25.1 shows, our third quarter profits rose by 12% overall, but our main services were down by 2%, 4%, and 1% respectively. This data indicates that our business is evolving and we should reconsider how we allocate resources to our various services."

Refer to the Reports chapter for more information on choosing the best visual and how to integrate the visual into your final report.

There is also more information on the report format for the Client Project under the Final Report section of this project.

Project Plan

Each team will be responsible for co-writing a project plan report memo and presenting it to the class as a brief oral progress report.

The project plan is a preliminary report that:

- Defines the team's project
- Articulates a plan for completing the project, including research
- Persuades the reader, your instructor, the project is well thought out and feasible.

The project plan answers the questions: How will you proceed to provide the client with relevant and useful information? Have you thought the client's need or problem thoroughly enough to proceed with the support of your instructor and the client? The project plan also helps the project participants themselves, for whom such a report acts as a guide, almost a contract, to specify who will do what and when. If done thoroughly enough it will also serve as a draft of your methodology section of the final report. For more information, refer back to the Establishing a Plan to Manage the Project part of the Planning section.

Your team's project plan should be written in memo format and include the following:

- **Client Description**: Repurpose your client description from the project proposal.
- **Client Problem**: Repurpose from the project proposal, unless the team has decided to refocus the problem in some way.
- **Methodology**: Describe the team's research methodology, which includes the main research questions the team will pursue, what methods (both primary and secondary) the team will employ, and a justification of the methods you've selected, i.e., why are you using these methods? Why are they appropriate given your client's problem?
- **Project Schedule**: Describe of the team's plan for completing the project, including a discussion of who is going to do what, by when. Also include major milestones like progress report and final report draft/final deadlines. You must include a **Gantt chart** visualizing the team's schedule (but remember, the visual alone is not enough; you must explain the visual in your prose discussion).
- **Closing:** Brief statement about status of project. If there are any questions, obstacles, or problems the team is facing, you can mention them (optional).

Oral Presentation of Project Plan

The oral report's purposes are to inform the instructor and other teams of your group's research design and project plan. After a brief review of the client and problem, focus on describing in some detail your methodology, or the research methods your group will use accompanied by a rationale or justification for those methods (i.e., detail both *what* methods you're using and *why* you're using those methods).

The oral report should be approximately 5-8 minutes and be supported by at least 2 visuals. One visual must be a **Gantt chart** depicting the team's production schedule or a **Research Design Chart** depicting the team's research design. The other visual can be whatever the team deems most appropriate to support their discussion, such as textual outlines of the oral progress update, diagrams of the client organization, or the team's analysis of the client's need/problem (possibly using the research design charts). Teams are free to choose the medium for visually supporting their reports (e.g., overhead transparencies, Word files, PowerPoint, etc.).

Client Letter

After your class has selected clients and formed project teams, your instructor may ask you to write a letter to your client informing him/her of the status of your proposal and thanking him/her for participating in your education.

For some, this will mean writing an acceptance letter, telling the client your proposal was among those selected and informing them of the next steps in the project (i.e., contacting the client to set up a meeting with the group).

For most others, the client letter will be a rejection letter, telling the client that their project was not among those selected by the class. The rejection letter is more challenging to write, because it must deliver "bad news" in a way that maintains good will.

Write a short (less than one page) acceptance/rejection letter in proper business letter format:

- Inform client of status of proposal
- In the case of rejection letters, try to politely indicate why letter wasn't selected (e.g., "…only 4 of 21 proposals were accepted")
- In case of acceptance letter, remind them next step is meeting with group to discuss client's ideas for meeting needs and develop suitable research plan
- Thank client for his or her time and interest
- Remember to maintain a professional tone (include a buffer, an indirect lead in, in the opening of rejection letters)

Sample letters?

There's no template for this assignment. You'll have to use your best judgment to craft your client letter. You can, however, think about adapting the basic formats and styles of acceptance and rejection letters to your purpose and audience.

- "Declining Job Offer" letter
  http://www.quintcareers.com/sample_declining_letter.html
- About.com's "Letter Writing Desk" includes essential resources, formats and downloadable samples for writing resignation, thank-you, reference (recommendation), salary and resume cover letters.
  htttp://jobsearchtech.about.com/library/bl-business-letters.htm

E-Mail Progress Report

Each team member may also be asked to submit an individual or group progress report individual progress reports inform the instructor of an individual's contribution to the overall team project. An individual report places one's own efforts in the context of the whole group's progress. A group progress focuses on the group's progress overall, but should discuss how individuals have been doing completing the individual research tasks. The report should be approximately 1,000 words, so maintaining clarity and concision will be important.

Organizing the Progress Report

Progress reports are common in business. The genre is designed to justify work done, evaluate that work, and also to forecast work remaining. As a genre, the progress report is conventionally divided into two major sections, work completed and work remaining. Context is provided by overview, background, and conclusion sections.

Include the following sections in your progress report. You are not limited to these sections; you may want to subdivide the sections, for instance, to report on different work completed or remaining:

- **Overview**: Briefly summarize of purpose and content of document. This report is focused on informing the reader of the status of the Client Project.
- **Background/Context**: Reiterate for the reader the purpose of research project, including client, client need/problem. Outline initial project plan, including project schedule, designated tasks and sources of information. Answers questions: What is the project? Is the project proceeding as planned? Have there been any significant changes in the scope or organization of the project?
- **Work Completed (or Results to Date)**: What data do you have to report at this point? What preliminary (tentative) conclusions can you offer? You should describe actual results; not just merely say, "I've made lots of progress." This reporting should resemble the "Methods" section of a final Recommendation Report, justifying your selection of sources and choice of research methods. Yet, like the findings and recommendation sections, it will also report on the significance of the data. It is the opportunity to articulate initial insights gained from the research.

    For example, concerning interviews, answer the following questions: What were your plans for conducting the personal interviews? Why were you doing them/what do you hope to accomplish with them? Who (name and title) in the organization have you decided to interview and why? What questions did you plan to ask? (For surveys, you may even attach a copy of your questions.) How long did it take? Are these one-time only interviews or do you plan a follow up as well? What were the results? What kind of notes did you take? What information was valuable? Why was it significant? How does this help answer your research questions or solve client's problem?

- **Work Remaining**: This section should describe explicitly tasks which remain to be completed by you and the group before the completion of the project. Considerations resemble the work completed section (i.e., justifying tasks, sources and methods) though there is no data as yet to report.
- **Conclusion**: Reiterates status of project and assesses the chances for its success. As a writer, you want to be persuasive as well as honest. Your tone should be positive, but if there are any concerns with the project, you should not avoid mentioning them. Your decision-making readers need to be given an honest report so that they can offer assistance, redirect resources, etc.

Final Report

The following information describes the features of your Final Recommendation Report and Oral Presentation for the Client Project. Business reports are further discussed in the Reports chapter of this textbook.

On the day the Final Recommendation Report is due, you'll need to prepare two (2) copies of the report, one for your instructor and one for the client. Your final written report must include all of the elements discussed below. Please pay close attention to the following throughout your planning and writing process.

- Goals for the Client Recommendation Report
- Report Format
- Review Checklist for Final Report
- The Oral Presentation

Goals for the Client Recommendation Report

Your overall aim is to help the client improve their business communication practices so that the organization can function more effectively and fulfill its goals. The client will use this report to help them decide what changes in their organization are necessary to make.

Analyze the Rhetorical Situation of the Client Report

Purposes

- Why are you writing this report? What are the main purposes and what are secondary purposes?
- What are your goals for this report? Which sections should be clearly related to the goals of the report?
- What different types of evidence are available to you?
- How can these be used effectively?

Writer(s)
- Who are the writers of the report? Are they experts?
- What is their relationship to the readers?

Reader(s)
- List all possible readers (both primary and secondary) and describe their relationship(s) to the material.

Text
- What form/design should the report take?
- How can this form/design help you achieve your goals?
- How many different ways can you arrange the report?
- What material will you choose to keep?
- What material will you choose to delete? Why?

Report Format

Consult the website companion for sample client reports and report templates

Each recommendation report should be between approximately 15-20 pages and **include at least two visuals**, one tables and one figures (see the Using Visuals in Your Report section of the Reports chapter). These page guidelines are just that—guidelines. Remember that professional quality business reports use appropriate page design, so page lengths may vary according to your particular project, how many visuals you use, how much white space you have in the margins and between sections, etc.

Ordinarily, students prepare two copies, one for teacher and one for client. Formal reports are usually bound. You should make an effort to have the report nicely bound by your campus copy center or some other commercial copy store, for appearance does make an impact in formal reports. (Your instructor may or may not require this.)

Your report must include the following elements in some form:

1. Front Matter of Report
    1.1. Letter of Transmittal
    1.2. Cover/Title Page
    1.3. Executive Summary
    1.4. Table of Contents

2. Major Sections of Report
   2.1. Introduction
   2.2. Background
   2.3. Recommendations
   2.4. Methods & Findings
   2.5. Conclusion
3. Back Matter of Report
   3.1. List of Sources
   3.2. Appendices

1. Front Matter of Report

1.1 Letter of transmittal

The letter of transmittal is a cover letter that delivers completed projects to external audiences. Generally, the letter should follow standard letter format and be limited to one page. These letters can either be bound with the report or delivered with the report as a separate document. The letter should contain the following elements:

- Opening Paragraph: statement of purpose, the date the work was commissioned, statement that the report has been delivered, and the title of the report.
- 2nd Paragraph: briefly summarize your recommendations and the pertinent findings from your research, including one or two significant details.
- 3rd Paragraph: briefly describe your research methods, as well as acknowledging any key contributors beyond the consulting group and important sources.
- 4th Paragraph: close courteously, express a willingness to work further with the client, offer to answer any questions, and provide an open avenue for communication.

1.2 Cover/Title Page

Generally, the cover includes names of the writers, title of report, name of recipient, and date of report. The cover page should have the UNLV Business Writing logo at the top of the page. Separate this information with a border. The title of the research comes next, followed by the date. At the bottom of the page, you need to list the following information: Instructor's name and UNLV address; the names of each member of the research team; the name of the client, client company, and company address.

How should you choose your title? The title should be brief, but must also convey something of the subject of the report to the reader. It's helpful to think about describing the purpose or objective of the report, such as "A Usability Study of MySportStat.com's Content" or "Web Advertising Strategies: A Recommendation Report."

1.3 Executive Summary/Informative abstract

An executive summary is a mini-version of the final report, designed for decision-making audiences who lack the time to read the entire report closely. This document should be written in lay terms, designed for managers who may not have the technical expertise of other readers. It should be no more than two (2) pages and is placed before the body of the report. Include the following parts:

- Clear description of the context that led to the research performed, including a discussion of the problem. Be sure to use specific terms and evidence to show the problem exists.
- Brief description of the methods used for research and the research findings would also be appropriate here.
- Detailed list of recommendations with some justification. A bulleted list of detailed recommendations is particularly effective.

1.4 Table of Contents

The table of contents provides readers with a way to find the information that they want quickly and efficiently. List all major headings and all subheadings.

Microsoft Word can generate a table of contents automatically using the INSERT> REFERENCE > INDEX AND TABLES menu. For this to work, all major headings and subheadings need to be formatted as headings. You can search Word's Help menu or find tutorials on the Internet for more instructions.

2.0 Major Sections of Report

2.1 Introduction

Like any opening, include a descriptive summary reviewing the gist of the report, focusing on the statement of the problem, pertinent background/history of project, and results/recommendations. Also include a summary of the report sections.

2.2 Background

Describe the context and purpose of the report in sufficient enough detail to justify report. Include:

- Client description
- Statement of client's need or problem
- Statement of project objectives

Be sure to convince the readers that there are possible problems by describing causes. Provide statistics, interview quotes, and any other evidence you can find to back up your points. Speculate about the problem/need: that is, if nothing is done, what will the future hold? Create a sense of urgency to convince the readers that they need to take action and implement the information that you have gathered. Remember that the primary audience is the client, so pay attention to the principles of maintaining goodwill and finding positive ways to express negative ideas when describing the client's problem (i.e., don't bash the client in the report, use more tactful language to discuss deficiencies in the client's organization).

2.3 Recommendations

Following the managerial organization pattern, the recommendations should come after the report introduction and background section (see Reports chapter). The recommendation section should:

- Give an answer to your guiding research question based on what the client will find feasible.
- Describe how the information you have gathered can best be implemented for the benefit of the organization.
- Convince your audience that your information is valid and important for the organization.

Make arguments following the general criteria for evaluating options listed below, and be sure to support your claims with evidence from your research

Effectiveness	• Will your recommended plan be effective? • Will it work? • Will the recommendations solve the problem?
Resource Feasibility	• Can client implement the recommendations? • Do the recommendations require technology or resources (e.g., personnel) that are unavailable? • What will have to be acquired to make the option work? • Will the recommendations require suspension of services or production? • Will there be any training involved? Hiring new personnel?
Desirability	• Will the organization want to implement the proposed options? • Will anyone have personal objections to the option? Do the recommendations harm or threaten anyone (either staff, clients, or customers)? • Is the option legal? Ethical? • Does the option have desirable effects? Potential undesirable effects • What are they?
Affordability	• What does the option cost to implement? To maintain? Is the cost reasonable? • Is the cost justifiable given the level of need or severity of the problem?
Preferability	• Is the option better than others? Why? • Develop and explain the criteria you use to make comparisons (e.g., choosing a computer system based on criteria such as: cost, processing speed, software package, ease-of-use, technical support)

Source: Leslie Olsen and Thomas Huckin. *Technical Writing and Professional Communication* (2nd ed., McGraw-Hill, 1991).

2.4 Methods and Findings

The methods and findings section, sometimes referred to as the methodology section, follows the recommendations section. It should include the following sections in your methodology section:

- **Findings** section that discusses the results of your research. List and provide details about what you have found out, including what your surveys, personal interviews, user testing and any secondary source material revealed (this can be a subsection of Methodology or a major section of its own).
- **Methods** section that describes and justifies your research techniques

Critical to any primary research report is the data that you collect. You present it in various tables, charts, and graphs (see the section on creating, formatting, and incorporating graphics into your reports). Call attention to relevant results (don't leave it up to reader). Less important findings can go in appendixes if they are so big that they interrupt the flow of your discussion.

2.5 Conclusion

Include a persuasive closing, convincing your audience to accept your study and recommendations. Suggest benefits of accepting report's claims.

3.0 Back Matter of Report

3.1 List of Sources

Cite your outside information/sources using an accepted style manual. Choose Chicago or APA style and follow it consistently by imitating examples for citing various sources.

3.2 Appendices

An appendix is a place to put information that just will not fit in the main body of the report but still needs to be in the report. Big tables of data, large maps, forms used in an organization, or verbatim transcripts from interviews—these are good candidates for an appendix. Each one is given a letter (Appendix A, B, C, and so on). You can include photocopies of related information. You can also include survey or interview results, but you should never include "raw" data; always present your results so that your reader does not have to sift through large amounts of information. Data should always be processed, condensed, and tabulated for the reader's benefit. Reader are free to disagree with the report writer's interpretation of the data, but the report writer and readers should never disagree on what data is being presented.

Remember that most charts and graphs should be placed as close as possible to the relevant discussion in the body of the report. You may not need to have appendices in your report if you only have visuals but not any supporting documentation to include.

Review Checklist for Final Report

Format

- Is the audience responsible for seeing that the findings are used and the recommendations implemented?
- Does the report use effective, informative headings? Are they significant of the content of the section?
- Is it easy to distinguish between main sections and sub-sections?
- Is white space used effectively? Does the page design look choppy or busy?
- Are the paragraphs and sections focused and persuasive?
- Has the writer used a professional tone?
- Is the report paginated?
- Does your executive summary review key points from each section of the report, informing readers of recommendations and conclusion?

Organization

- Are the sections arranged most logically and effectively for understanding the information and its relationship to the purpose and thesis of the report?
- Does the report have an introduction, which includes the purpose and forecast of the document and conclusion?
- Does the background section summarize the problem/situation?
- In what ways does each section support or relate to the main point?
- Are the research methods clearly explained?
- Does the conclusion restate the main argument of the report? Does it indicate what the audience should do with the findings? Does it provide contact information from the writers?
- Does the conclusion create a sense of urgency?
- If there are appendices, is a list of their titles provided immediately following the conclusion?

Findings

- Are the findings presented as conclusions about specific issues?
- Are results used to support, contradict, or qualify other findings?
- Are tables or figures used to clarify complex data?

Recommendations

- Are the recommendations explained in persuasive detail?
- Do the recommendations meet the needs of the organization?
- Are the recommendations feasible in terms of the organization's constraints and goals?
- Are the recommendations supported with findings?
- Does the discussion provide a brief suggestion of how the recommendations could/should be implemented?

Tables or Figures

- Have you included at least 2 visual/graphic aids?
- Are your graphics consistently formatted (e.g., have titles and labels) and appropriated placed?
- Do the graphs visually support the discussion of them? ? Are their titles indicative of their content?
- Is each visual the most effective in representing the data?
- Are other graphs needed to make a discussion more clear?
- Is the interpretation of the graphs in the discussion well developed and clear so that graphs would be unnecessary?
- What results do the graphs reveal that the writers did not discuss?

Sources

- Are you citing sources using a specific style guide like APA or Chicago Style? This includes in-text citations (making references whenever you're quoting, paraphrasing, or summarizing) and the list of references at the end of the report. URL: http://library.nevada.edu/subjects/style.html

Project Assessment Memo

In a formal business e-mail memo no longer than 500 words, evaluate the success of your project, the final versions of documents, and how your team performed as a group.

This PAM is primarily a group performance evaluation. Your main purpose is to inform your instructor of how your group performed throughout the project and whether or not there were any problems that affected the outcome of the project. Remember, each individual receives a separate grade for this project, and your instructor will use the final project evaluation, your PAMs, and his or her familiarity with your group to assess final grades for each team member. Team members do not automatically receive the same grade. Poor participation could hurt your grade.

Be sure to assess the performance of each team member, including yourself, according to the following criteria:

- Contributed his or her fair share to all phases of the project?
- Participated actively in meetings?
- Was dependable, prompt, and courteous as a group member?

For each team member, including yourself, provide an overall rating for his or her contributions to the group. You could use a scale of 1 to 5 (be sure to identify which is best) or assign letter grades.

If you feel one or more persons did not contribute equally, be sure to provide concrete reasons why they didn't contribute. Likewise, if you feel some members of your group are dissatisfied with your performance, you might want to provide an explanation from your point of view.

Don't forget to comment on the quality of your final project.

Chapter 26
Usability Project

> **Project Objectives**
>
> - Practice writing proposals, progress reports, and technical research reports
> - Apply project management, research, and team writing techniques to the production of effective reports
> - Plan and carry out a multi-stage, collaborative research and writing project
> - Research, analyze and present large amounts of information clearly and persuasively in professional quality reports
> - Understand concepts of usability and user testing in effective document design
> - Identify manageable problems or needs within organizations and articulate them as researchable questions

For this major collaborative research project, you will work in teams of 3 to 4 to evaluate an existing Web site that belongs to an actual person or organization. You will assume the role of a usability consultant and act as if an actual client, selected by your team, wishes to improve his or her organization's Web site to meet the needs of a specific target audience. As a usability consultant, you are committed to the idea of user-centered design—that is, designing documents by focusing on potential users from the very beginning, and checking at each step of the way with these users to ensure that they will like and be comfortable with the final product.

Your job will be to help a client, whom you may or may not actually interact with, by conducting primary research about the Web site and providing feedback and design advice for a more user-centered Web site. For this project, then, you will:

- Identify an actual client organization that has an interactive Web site
- Describe in detail the client organization and target demographic for the Web site. The more information that you can begin with, the better your research design and analysis will be
- Design a research plan to elicit relevant information for your client about the usability of its Web site

- Perform tests on the existing site with intended users to determine its effectiveness in specific areas
- Analyze the extent to which your client's site achieves its purposes as well as those of the users
- Write a usability recommendation report addressed to the client organization that (1) provides the client with information about your assessment of the users' needs and the usability of the site, and (2) recommends ways the client might revise, upgrade, or improve its web-based communication strategy

Each group articulates the results of its evaluation in an analytical report addressed to the client. This project introduces students to the *report writing cycle*, or the process of documents and deadlines by which collaboratively written business reports are produced. Students write a variety of documents, including a project plan and progress report, culminating in the final report. Students also submit individual team performance evaluations that assess the contributions of each team member.

What Does Usability Have to do with Learning How to Write?

This project familiarizes you with collaborative report writing techniques by asking you to act as part of a Web design consulting team. Each team will research targeted groups of users and, based on this information, make recommendations about how to upgrade Web-based documents to better perform for these target groups. Usability consulting is the shorthand for this service. Refer to the User Testing chapter for an overview of user testing.

The essential idea of usability, that the effectiveness of important documents can be improved with input from actual intended users, is older than web-based documents and can be applied to the production of any document, print or online.

While most of you won't go into careers as web designers, knowing how to identify good, usable web design will help you produce better documents. Like all effective business writing, a good Web site takes into account rhetorical issues of audience and purpose. Because the Internet allows for the presentation of information in multi-media, issues of visual design are critical.

By acting as a usability consultant, you will also have to apply all you've learned in business writing to produce your final recommendation report, including how to plan carefully, research thoroughly, think critically, and communicate effectively. Because you'll be working in a team—which mirrors how most reports are written in the workplace—you'll also have to utilize teamwork, project management, and group writing skills.

Deliverables

Document	Writer(s)	Audience	Format
1. Design Plan Report	Group	Instructor	3–4 pages; 11 pt font, single spaced; Must attach copies of user test protocol (data collection sheets)
2. Progress Report	Group	Instructor	E-mail; 850 words. Informs instructor of findings from user tests and status of project
3. Final Technical Report	Group	Client	12–15 pages (following good page design and layout); 12 pt. single spaced, plus appendices and references. Must include at least 3 visuals (incl. 1 table, 1 chart, and 1 graphic, usually a screenshot). Teams must write a report draft approved by instructor beforehand.
4. PAM	Individual	Instructor	E-mail; <500 words. Focuses on evaluating self and team members. Failure to complete equals letter grade reduction to final report grade

Choosing a Client Web Site

To prepare for usability project team selection, brainstorm a list of three possible clients. Keep in mind the following criteria for choosing a good client:

- **Can be a national, local, or university organization.** Examples of past student projects that involved national organizations include Ebay.com, MySportStat.com, or Fandango.com. Local organizations include area businesses or nonprofits, like a relative's accounting or law firm, or an animal shelter or youth foundation. University organizations are good choices, like UNLV college and department Web sites. Most student services, clubs, and organizations have Web sites too.
- **Must be an interactive Web site.** The Web site must involve interactive components that require the user to actually do something on the site, like move from page to page to find information. Most Web sites have specific functions too, like ticket sales, book listings, and airline reservations. A basic Web site with just one long, scrolling page of information, which requires little user interactivity, is a bad choice.
- **The more authentic and more possibility for client interaction, the better.** You should consider choosing a client that you have some existing relationship with, so you can talk to them about the Web site and your

project. This usually makes for a better, more meaningful project. Your initial analysis is also made easier because you're getting this information firsthand from the client rather than having to estimate these elements yourself. Maybe your mom is trying to sell her cookies online (an actual student project) or your friend has a Web site promoting the Las Vegas music scene (another actual project).

- **Should target class demographic.** You want to choose a client that lets you use classmates as the test subjects who provide user feedback about your Web site. The project is set up so that you will have time in class to work on planning your project, including conducting the actual Web site evaluation/usability tests. Obviously, if you want to test intended users in class, then they should already be represented in the typical business writing class. It might be hard to do in-class testing on a medical Web site whose target audience is senior citizens, but it will be easy to test the university's Registrar Web site. (You are not prohibited from testing out of class. Students often elect to conduct more tests than in-class time allows.)

Think twice about choosing a high end, professionally produced Web site. A good user test can reveal strengths and weaknesses to any Web site. But since you're not required to be a usability expert for this project, you might want to avoid popular Web sites that have probably already been extensively user tested. The more popular the Web site, the more you'll have to know about Web design and usability to offer substantive recommendations. Think about Web sites you've visited recently and think if you're aware of any usability problems from your own experience using these sites.

> **Online Companion**
>
> Use the Design Plan Worksheet to plan your project and write your Design Plan.

Design Plan Report

In a memo 4 pages or less, not including attachments, your team must inform your instructor of the following:

- Your choice of Web site for this project
- Your initial analysis of the Web site
- Your plan for evaluating the Web site
- Your group's schedule for completing the project

One of your primary purposes is to persuade your instructor that you have a well thought out plan for completing the research phase of your Usability Project. You need to argue your plan for evaluating the Web site. Your Design Plan should clearly communicate your goals, present an initial analysis of the Web site, and articulate your plan for conducting the user tests.

The Design Plan is an important first step in the report writing cycle. By writing this planning document—a form of a proposal—your team is articulating its objectives, which will serve as a blueprint for completing the project. It's much easier to successfully complete a project after you've articulated what your goals are, what needs to get done, who's going to do it, by when. You can use your plan to monitor the team's progress and stay aware of deadlines. By putting the plan in writing, the Design Plan also includes information that can be re-used later in the final report. This practice of using text from one document for other documents is called boilerplatting and is commonplace in business writing.

Format

Your Design Plan must include the following information:

- **Opening**: As with all business correspondence, open your memo with a brief statement of purpose and summary of main points. Do not use a heading for the opening section.
- **Background:** Who is your client, i.e., what is the Web site? Discuss the site's primary and secondary functions/purposes and its target audience/user groups. For example, a travel site's primary purpose may be to book travel reservations. Its secondary purposes might include providing destination information and instructions on booking reservations online. The people that would use this site include Internet savvy adults that prefer to make their own travel plans. They likely comparison shop at other travel sites. Most have probably booked travel online, etc. Consider if your Web site has multiple audience. For example, user groups for university sites include prospective students, current students (who have different needs than prospective students), visitors, athletic boosters, alumni, faculty, etc.
- **Preliminary Evaluation**: What is your initial evaluation of the web site? What are the strengths and weaknesses of the site's navigation, content, visual design, and functionality? Give reasons to support your evaluation. To help you derive some basis for analysis, include comparisons of your Web site to similar/competitor sites (e.g., "Compared to competitor sites X and Y, we believe the site…."). Comparing the client site to competitor sites is an excellent way to discover the site's strengths and weaknesses.

    NOTE: You might want to consider adding a more formal comparison to other Web sites as part of your method. In this approach, you would develop a more detailed comparison to one or two sites in your final report's findings section, which will allow you to base some of your recommendations on not only user test results, but also your own analysis of competitor sites.

- **Testing Goals**: What areas of the Web site will you evaluate? This should follow from your preliminary evaluation. For example, if an analysis of a business Web site reveals that it is (1) difficult to navigate, (2) hard to find information about the company's services, and (3) hard to read, you might articulate the following project goals:

  We will evaluate the following areas:

  1. *Navigation--Can the user browse the Web site without complication?*

  2. *Content--Can users find specific information regarding the company?*

  3. *Readability--Is the text size and background color making the information hard to read?*

  As the example above shows, state the project goals as research questions. This will make it easier in the Method section to create specific tasks that address the areas you aim to study.

- **Method**: Describe your team's plans user testing. You must attach a draft of your user test protocol/data collection sheets so your instructor can review them. In this section, be very detailed in your description of test procedures: How many users will you test? (NOTE: testing 5-10 users is sufficient for usability analysis.) What tasks will you ask them to perform? What data will you collect (before, during, and after the tests)? Discuss your procedure for conducting the tests. Will you administer a written protocol? Will it be interview-based? See the Usability Testing chapter for instructions on planning your test methods.

  If your team has elected to include a comparative analysis as part of the method, be sure to also discuss in this section (1) which sites you will be comparing and why you chose these sites, and (2) the criteria you will be applying in your analysis (see, e.g., "Web Design Criteria" http://www.frederick.k12.va.us/wq/webdsgn/process.htm).

  Briefly describe why the methods you chose are appropriate. Why are you asking certain questions in the pre-test portion of the user test? One reason, for example, is that you want to verify that the people you test match the typical user profile of the Web site. Or another example, you plan on collecting information about people's online shopping habits because you want to gauge their familiarity with using e-commerce Web sites. Why are the tasks you plan to administer appropriate? For example, the direct tasks of your test will allow you to determine how difficult it is for users to access information, complete a purchase, register for a login account, etc. Why are you only testing 5 to 10 people? (Jackob Nielsen writes that testing just five users can identify 80% of major usability problems.)

- **Schedule**: Describe your plan for managing the project. Include description of discrete tasks, individual/team responsibilities, time estimates, and major milestones/deadlines. Be sure to note key project milestones: When will you administer your tests, collate the information, analyze your data, draft your reports, and submit the final version? Illustrate your schedule discussion with a Gantt chart. (For a template, download the Creating Gantt Charts file from this textbook's companion Web site.) Remember, the Gantt chart alone is not sufficient—you must reference it and explain it in a prose discussion of your schedule.

- **Conclusion:** Conclude with an argument about the doability your project. Make sure your design plan is reasonable for the scope and timeframe of the project.
- **Attachment**

    Attach a copy of your team's test protocol, i.e., any questionnaires you plan to distribute to users during the tests and any other data collection sheets you plan to use.

Design Plan Review Checklist

- Is the Design Plan written in memo format, with proper heading, opening, body, closing, and end notation?
- Does the Design Plan follow the organization and heading-subheading structure outlined above?
- Has the Web site's purposes and target audience been described in detail?
- Does the design plan offer a preliminary analysis of the Web sites strengths and weaknesses? Is the preliminary analysis derived in part from comparison to other similar/competitor Web sites?
- Are the group's testing goals clear? What does the group want to learn from the testing about the user, the web site, the way the Web site is used/read, etc.?
- What methods will be used to meet the testing goals?
    - How will test subjects be selected?
    - When and where will the testing and/or interviews be conducted?
    - What specific tasks will the users be given? Is there a protocol for the test(s)?
    - What materials will be needed? (e.g., recording devices, product, documents, etc.)
    - Who will do what for the research and the testing?
- Is there a discussion of the group's schedule and does it reference a Gantt chart illustrating major tasks, milestones, and timelines?
- Is your team's user test protocol/data collection sheet attached?

Progress Report

Each team is responsible for submitting a group-written progress report that presents the results of the usability tests. It should be written in memo format, not exceed four pages, and include at least two visuals (table, graph, chart, or picture/screenshot).

Organizing the Progress Report

Remember that a progress report informs readers of the status of a project. It is designed to justify work done, evaluate that work, and also to forecast work remaining.

As a genre, the progress report focuses on work completed and work remaining—with context provided by overview, background, and conclusion sections. As with all effectively organized reports, you are not limited to these sections. For example, you could use the types of research completed as subsection headings.

- **Overview**: Open your report with a brief summary of purpose and content of document.
- **Background**: The next section reiterates for the reader the purpose of the research project. Summarize what Web site your team selected and your project objectives. (*Hint*: Boilerplate this information from your original Design Plan Report.) Also, if there have been any significant changes in the scope or organization of the project, you should note them.
- **Work Completed**: This section describes actual results to date. For this project, you should be summarizing the results of your user tests. Compile your data and present it in a clear and readable fashion. Just writing "we've made lots of progress" without a detailed discussion to support your claim would not convince a reader that your group is progressing.

  For the usability project, this section should resemble the findings section of the final report by summarizing the results of your user tests. You must include at least two visuals in this section. Consider the best way to present your data graphically now, so you can re-use any visual aids in your final report. Will a table work? Should you use a pie or bar chart? How will you summarize and present the results of any open-ended questions you asked? Is there a way to present this information visually (e.g., categorizing and tallying types of answers in a table). Or will you just summarize results and quote exemplary answers?

  Another useful visual aid when it comes to analyzing Web sites is the screen shot, literally an image of the computer screen that shows the relevant part of the Web site captured using the Print Screen key. As Seth Gordon recommends, "highlight problematic areas through a combination of narrative and screen captures. The screen shots are especially effective for communicating the context of the problem. It's often easy to see the problem but extremely difficult to document it." Gordon also suggests mentioning significant positive and negative feedback and any interesting or unexpected findings when discussing results of the user tests. Finally, according to Gordon, "a good evaluation captures respondent feedback, and the best way to communicate that is through respondent quotes"[1]

- **Work Remaining**: This section should describe tasks that remain to be completed by the group before the completion of the project. For instance, what type of recommendations does the group foresee making? How have the test results impacted your plans for the report? What work have you done towards drafting the report so far? And what work, specifically, remains to be done?
- **Conclusion**: Your final section should briefly reiterate the status of your project and assess the chances for its success. You want to be persuasive as well as honest. Your tone should be positive, but if there are any concerns with the project, you should not avoid mentioning them. Your reader, your teacher or manager, needs to be given an honest report so that s/he can offer assistance, re-direct resources, etc.

Work Cited

Gordon, Seth. "User Testing: Reporting Your Findings." *Builder.com* 6 Oct.2003
   http://articles.techrepublic.com.com/5100-22_11-5080692.html

Usability Report

Refer to the Reports chapter of this textbook for more information on formatting business reports. The usability report is its own sub-category, or genre, of business report and a simple Internet search should locate several examples. You can use these examples as guides, but keep in mind that not every usability report presents information in the clearest and most effective way.

The main sections of the report should be as follows:

- Overview
- Recommendations
- User Test Results
- Conclusion

Don't forget to also include the following conventional elements of reports:

- Front matter
    - Letter of transmittal
    - Cover
    - Executive Summary
    - Table of Contents
    - List of Figures
- Back matter
    - References (if applicable)
    - Appendices (if applicable)

Usability report recommendations are often written in a problem/solution format, whereby each problem area identified by the user testing is noted (with concrete references to user test data as evidence) and then suggestions or recommendations for addressing the problem area are given.

Solutions to most usability problems are usually fairly obvious, once an actual user identifies something as a problem. Solution ideas might come from user feedback/suggestions or a comparison to other Web sites, particularly competitor sites. Since you are not expected to be a usability expert for this project, your report should include recommendations for improving weak areas but will not necessarily be evaluated on the technical effectiveness of the solution.

As the Progress Report section in this project and the chapter on Reports indicates, screenshots are almost essential when conveying the results and recommendations of your user tests to a reader (see Figure 26.2). A screenshot, or screen capture, is an image of whatever is displayed on a computer monitor at the time of the screen capture. *Callouts*, or textboxes with arrows drawn to parts of the screenshot, help clarify what is being illustrated in the screenshot. Screenshots are useful for pointing to flaws in the client website or pointing out other Web sites that have better design elements. The Report chapter has more information on how to take screenshots and insert them into your document.

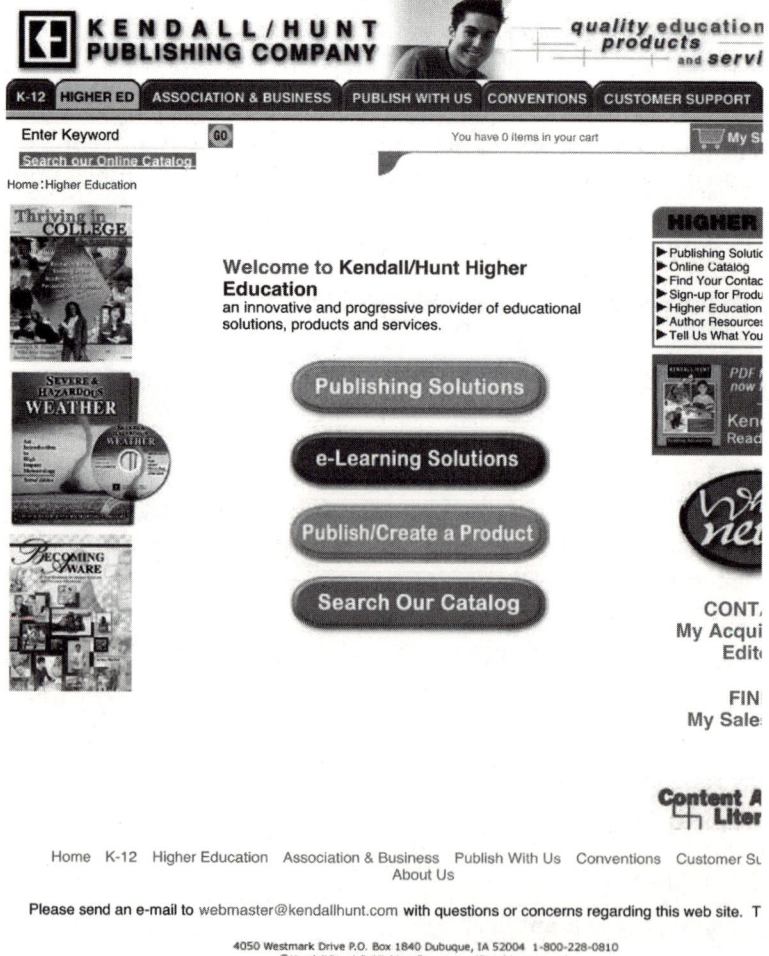

Figure 26.2. *Sample excerpt from a student usability report with screenshot, callout, and discussion of visual in body of the report.*

Usability Report Evaluation Checklist

Refer to the following checklists as your prepare your final report:

Formatting
- Are all report parts included: cover, transmittal letter, table of contents, list of figures, executive abstract, body, references (optional), appendices (optional)?
- Does the report use frequent, effective headings and subheadings, and is it easy to distinguish between main sections and sub-sections?
- Are headings (L1, L2, L3, L4) formatted consistently from section to section and from front matter to body?
- Is the report paginated following guidelines for page numbering, e.g., lower-case Roman numerals in front matter, Arabic numerals in body?
- Does the same header and footer consistently appear on all pages of report from front matter to body, except for letter of transmittal?
- Do you have a List of Figures page separate from the table of contents?
- Is white space used effectively, or does the page design look choppy, blank, or busy? Make sure format (e.g., headings, header/footers) is same for front matter as report proper

Recommendations
- Are recommendations subdivided with "talking" subheadings (i.e., starting heading with an action verb, e.g. "Recommendation #1: Minimize busy graphics...")
- Are recommendations developed in sufficient detail including (1) support from references to user test results, comparison to other Web sites, and/or expert opinion, and (2) suggestions for improving the problems, including possible screenshot illustrations?
- Do the recommendations in Executive Summary correspond to the recommendations in the main body of report?

Background and Methods
- Does background and method information correspond to content from the design plan, i.e., did you boilerplate your design plan content (minus the Schedule section) into the report?
- Is the "Method" section written in past tense (e.g., we asked users to...)

Graphics
- Is the minimum three graphics included (one chart, one table, and one screenshot)?
- Are all graphics labeled with a caption, including a consecutive number and clear title that conveys the main point of the visual?

- Are all graphics referenced in the text (e.g., "As Figure 26.1 shows,...")?
- Are all graphics placed as close as possible after the initial reference in the text? (Don't place graphic before; if it can't fit on the page that it is first mentioned, but it at the start of the next page and let reader know: "see page #.")
- Is the graphic simple, uncluttered, and easy to read, i.e., is visual unclear, either because of design of visual, inappropriate visual choice, or trying to convey too much information in one visual?

Style and Editing
- Does the report follow a more formal style than business letters and memos, e.g., does it avoid references to the reader as "you" (except in transmittal letter, where you should use it)? Are informal, slangy word choices avoided, such as "things" and "report is broken down by"?
- Are pronouns used appropriately and consistently? Don't use *you* as a third-person reference, write *users* or *one* instead. Check for pronoun agreement, e.g., "we asked one user what they thought" is incorrect because the subject *one* is singular and the pronoun *they* is plural.
- Are words spelled consistently, e.g., how did your team decide to spell W-E-B-S-I-T-E?
- Have you eliminated *telegraphic writing*, where words such as articles (a, an, the) are missing or the tense is wrong?
- Are there no stacked headings, i.e., does the report flow from one section to another when read aloud, or do more transitions from section to section need to be added so the reader doesn't get confused (e.g., "In this section, we discuss the methods we used...")?

Project Assessment Memo

See the instructions for writing the **PAM for the Client Project**. As with that project, the PAM for the Usability Project should be an evaluation of your group's members. In a formal business e-mail memo no longer than 500 words, evaluate the success of your project, the final versions of documents, and how your team performed as a group. Remember to speak specifically about how each member contribute to the project, including yourself. If here is any reason you believe someone should not receive the same grade as other students, you should mention it and provide reasons for your assertion. Likewise, if you believe other members in your group might single you out for some reason, you should indicate why you believe other team members might rate your contribution low.

Chapter 27

International Project

Project Objectives
- Learn the business report genre
- Practice the report writing cycle
- Understand and apply principles of project management
- Research and integrate information into business reports
- Evaluate data and information based on the needs of specific audiences
- Develop an understanding of cross-cultural communication

The International Project puts you and your group in a situation as researchers who will offer recommendations to the Bellcom corporation so that they may create materials to better orient their employees to the customs and features of their new international surroundings. To fulfill the requirements for this project, you will need to recommend the kinds of materials for a complete package that prepares employees for the move.

Our class will take on the persona of a consulting group, a company which provides a wide variety of research services. Our most recent client, the Bellcom Corporation, is expanding its services internationally, and has had trouble recently with employees traveling, living, and conducting business abroad for the first time (see chapter 18, A Business Faux Pas Case). At this time, Bellcom is opening new offices in the following cities:

- Amsterdam , Netherlands
- Athens , Greece
- Brussels , Belgium
- Dublin , Ireland
- Mexico City , Mexico
- Nairobi , Kenya
- New Delhi , India
- Rio de Janeiro , Brazil

While these sites are ideally suited for their business purposes, Bellcom wants to prepare the employees who will be working in these cities for the kinds of cultural differences they will face. Our research company has been hired to make recommendations about the kinds of materials that Bellcom should have developed to aid their employees preparing to move to their new surroundings. Your primary contact at Bellcom for this research is Sharika Jones, the vice president of human resources.

Your major collaborative project in this class will be to collect information so that you can make recommendations for developing a travel guide. While Bellcom wants to be comprehensive in developing their travel guides, they also want to make sure that they are not offering unnecessary or redundant information. Your goal is to create a recommendation report that describes and argues for the most important information that employees will need upon moving to a particular city. By performing this assignment, you will develop skills that are valued in the business world and, therefore, provide you with certifiable experience.

Deliverables

This project will include the following documents:

- **Research Design Plan**: 3–5 page memo informing instructor of your plan for effectively completing the project
- **Progress Report**: 3–5 pages e-memo informing your instructor of your initial findings, analyses, and work to be completed
- **Final Report**: 8–12 pages (good page design and layout) report addressed to Sharika Jones of Bellcom (include appendices, references, and at least three visuals)
- **Project Assessment Memo**: 500-word e-mail evaluating self, team, and overall project. Failure to complete equals letter grade reduction to final report grade

---

Readings

For the International Project, you should read the following selections in this textbook:
- Usability Project Design Plan
- Usability Project Progress Report
- Client Project Final Report

While the above readings are not direct correlations to the kind of writing that we are doing for this project, they will give you a rough overview of the kinds of documents that you need to create.

Planning

Each group will be responsible for the following in order to complete this project:

- Determine your goals as a group
- Decide how your group will accomplish the work
- Create a research agenda
- Perform research and evaluate information
- Develop most appropriate recommendations for situation
- Construct, evaluate, and revise your report
- Organize your report so that information can be found quickly and easily

Project Planning

To begin evaluating the material requested by our client, you will have to plan and set up appropriate research activities. You should distribute the workload evenly, and each member of the group should participate in all phases of document development. To gather good-quality information, you need to carefully avoid setting up activities that give irrelevant or even misleading information; therefore, as a group you should develop criteria that will assist you in analyzing and evaluating information so that you can gather the best information possible. To help avoid research problems, you should consider the following:

- who the client is and what kind of information they need
- the strengths and weaknesses of each team member
- main topics for research
- appropriate criteria for evaluating information
- research sources and strategies
- potential obstacles or problems in gathering information
- a Gantt chart or other organizational document to insure that the group stays on task

Research

You'll need to determine the most important information that Bellcom employees (most of whom have families) will need to know.

- What kinds of information can best help Americans adjust to the differences in culture? How do you define cultural information? Social/Political/Economic? More specifically?
- What are the most important pieces of information?
- What are the best sources of information?
- What are common stereotypes about the country? What values do these stereotypes suggest about the country? How can these be counteracted?

Some possible topics include:

- preparing to move
- general information about the city
- weather
- single-life/family life
- housing
- money
- transportation
- food
- health care
- schools/education
- entertainment
- culture and cultural differences
- language .

Document Production

Purpose(s)

- What are the purposes for writing?
- What are your/our goals?
- How much is enough information to make a credible argument?
- Does the information anticipate and answer the audience's questions? Why or why not?

Readers

- Who are the primary readers?
- Who are possible secondary readers?
- Describe your potential readers in as much detail as possible and explore the ways these readers will effect the documents that you are creating?

Writer

- Is the writer's identity important? Why or why not?

Text

- What form/design should the report take?
- What style/tone would be the most appropriate?
- How can this form/design help you achieve your goals?
- How many different ways can you arrange the materials?

- How can the information be organized most effectively?
- What kinds of graphics/graphs/pictures can you provide? How are these items helpful?

General Evaluation Criteria for the International Project

- When evaluating the documents that you create for this project, here are some general criteria for you to consider:
- Is there sufficient information?
- Will the reader feel they can make informed decisions in helping their employees prepare to move to this country?
- Are the cultural elements defined?
- Is the style/tone consistent?
- Are the documents easy to read?
- Is the information arranged effectively?
- Does the material provide sources for more information?
- Is the material arranged hierarchically?